Edward Atkinson

The Distribution of Products

Or, The Mechanism and the Metaphysics of Exchange. Third Edition

Edward Atkinson

The Distribution of Products
Or, The Mechanism and the Metaphysics of Exchange. Third Edition

ISBN/EAN: 9783337117658

Printed in Europe, USA, Canada, Australia, Japan

Cover: Foto ©ninafisch / pixelio.de

More available books at **www.hansebooks.com**

THE

DISTRIBUTION OF PRODUCTS

OR

THE MECHANISM AND THE METAPHYSICS
OF EXCHANGE

THREE ESSAYS

WHAT MAKES THE RATE OF WAGES?
WHAT IS A BANK?
THE RAILWAY, THE FARMER, AND THE PUBLIC

BY

EDWARD ATKINSON

THIRD EDITION

NEW YORK & LONDON
G. P. PUTNAM'S SONS
The Knickerbocker Press
1888

GENERAL PREFACE.

It may happen that one whose life from very early years has been of necessity mainly devoted to active business and to practical affairs, will be found as well qualified to treat the momentous questions which are the subjects of the following essays as students or practised writers whose pursuits are far removed from the actual work of providing for the material wants of men. But if this do not prove to be the case, yet the business man who puts the results of his observations into a simple form, easy of comprehension, may yet aid those who are more competent than himself in evolving a knowledge of the higher laws upon which the very existence of society depends.

A true commercial and economic history of nations, or even of the English-speaking people, remains to be written. "How did these people get their living?" is the question which every practical man asks when reading about the struggles of dynasties, the narrative of wars and battles, and the records of debates of legislative bodies, which constitute the chief material of history. Even when he reads such history with intelligent comprehension, he cannot fail to observe that no matter how each great struggle has begun, whether incited by religious enthusiasm, by personal ambition, or by the uprising of an oppressed people, in the end it has almost always been the commissariat that has controlled events. Power has fallen not so much to the strongest battalions and to the heaviest guns, as to those who could sustain the battalions longest, and support them with bread and meat as well as with powder and iron.

On the other hand, one who reads even the history of our own country from the commercial standpoint may well believe that had Adam Smith's "Wealth of Nations" been written fifty years earlier, it might have exercised as profound an effect on the commercial policy of England during the fifty years preceding 1776 as it did in the fifty years subsequent to that date, in which case the colonies of America might have separated from the mother country by peaceful methods, and the War of the Revolution might have been spared.

Only the present can be called a specifically commercial century, and one of its phases has been the abuse of the power of credit. Only in a commercial era could national debts have been incurred in the way they have been during the last fifty years, and now these debts threaten the very existence of the nations which are burthened by them. Since the beginning of the present century, the public debt of Europe has risen from $2,600,000,000 to over $22,000,000,000. This debt has been accompanied in many States by the issue of paper substitutes for money, which have depreciated, and either by that method, or in some more summary way, the repudiation of a large part of it may become a necessity before the end of the century.

There are two kinds of national debt. One consists of debts imposed upon the property and products of the people by a dynasty, or by a privileged class of legislators, without the consent of the governed, and generally for the prosecution of wars by which the people were oppressed rather than made free. So long as such a debt exists it works a false distribution of wealth and of product, and it has even been said by one of the greatest living statesmen of England, that her "national debt is the chief cause of her pauperism."

The other kind of national debt is one incurred by the consent of the governed for the purpose of establishing personal liberty and equal rights. Such is our debt. It will all, or nearly all, be paid within one generation from the date when it was incurred, or at least within the present century; and it will have fallen to a democratic nation, founded upon manhood suffrage, to be the first among nations to redeem a substitute for money, put into use as money under an act of legal tender for the purpose of collecting a forced loan, in the true coined money named in the promise. We shall also be the first to pay our debt without discount or depreciation. When our faith in democracy fails us, let us think of this and again take courage.

But if I can add nothing to the science of history by putting these practical treatises within the reach of students, I may yet offer them to my business friends and associates with the assurance that even if they may serve no other purpose, such studies lighten the necessary drudgery of our daily life, lend a phase of imagination to our work, bring friends and sympathy among men of science and literature, and render life far better worth living than it would be if it could only be measured by the mere dollars which we earn.

To the Directors of the Corporations with which I am now connected and by whom my own daily work is supervised and aided, I dedicate this volume, in testimony of the cordial friendship by which business men may become united, even when differing widely in their views upon great questions of public policy.

EDWARD ATKINSON.

BROOKLINE, *Oct.* 4, 1884.

WHAT MAKES THE RATE OF WAGES?

A TREATISE PREPARED BY

EDWARD ATKINSON
Of Boston, Mass., U. S. A.

AND SUBMITTED AT THE MEETING OF THE BRITISH ASSOCIATION FOR THE ADVANCEMENT OF SCIENCE, HELD IN MONTREAL, CANADA, AUGUST 28, 1884; ALSO PRESENTED BY TITLE AT THE MEETING OF THE AMERICAN SOCIAL SCIENCE ASSOCIATION, HELD IN SARATOGA, SEPTEMBER, 1884

"In proportion to the increase of capital the absolute share (of a given product) falling to capital is augmented, but the relative share is diminished; on the other hand, the share falling to labor is increased, both absolutely and relatively."—BASTIAT.

PREFACE.

THE only time which the writer could devote to the dictation of this treatise and to the computations which have been necessary in its preparation, has been in the short intervals of active business, and in the few evenings which could be spared after the duties of the day were over.

The treatise therefore takes a somewhat unsuitable form, consisting of introduction, the treatise proper, notes and explanations which have been added, and the various appendices sustaining the main argument,—many of which addenda are entitled to more consideration in the United States than they would have on the part of the British Association for the Advancement of Science, before whom the treatise proper was first read.

If time sufficed, all these detached portions might well be re-written and condensed in the treatise itself. But this is impossible. I submit the essay with the hope that it will give a direction to a thorough and complete official investigation, if the method is found to be a suitable one; or to a continuation of the study on the part of competent economists who have more time than I have to devote to such work, rather than with any expectation of its being accepted as final and conclusive.

All persons with whom I have conferred agree upon the paramount importance of the main question presented.

All men who have studied the phenomena of wages are somewhat appalled by the indications of the contest which seems to be approaching in every civilized state. This struggle takes the aspect in one place of a contest between the landlord and tenant ; in another between landowner and peasant ; in another between mill-owner and operative ; in another between privileged class and proletariat ; in another between rich and poor ; and in another between the needy and the well-to-do, whether the latter be rich or only well-off.

These are all different phases of the same question; all rest for their conclusion upon the simple problem: "WHAT MAKES THE RATE OF WAGES?"

Entirely subordinate to these great divisions between classes may be found the minor questions of State interference with the hours of labor; State regulation of the railway service ; protection and free trade ; usury laws ; the employment of women in factories ; and other labor questions so-called. In this treatise I have avoided, as far as possible, any reference to these minor questions in order to keep its main purpose distinct and separate.

I trust that an attempt to present, if not to determine, a fundamental principle underlying all these various questions will be as welcome to the advocate of protection as it may be to the advocate of free trade ; as welcome to the believer in coöperation, as it may be to one who trusts in competition ; as welcome to the person to whom the wrongs of the poor seem most urgent, as it may be to the man of wealth who considers his property a trust—involving duties as well as rights.

EDWARD ATKINSON.

BROOKLINE, MASS., U. S. A.,
September 18, 1884.

INTRODUCTION.

THE purpose of the following treatise is to consider the forces to which both employer and employed are subjected in determining what rates of wages can be paid in money and which control the bargains made between them.

It is not denied that an employer who is in the possession of large capital may agree to pay a certain rate of wages for a time, irrespective of any other conditions than his own will. But his power to do so will be limited by the amount of capital previously earned which he is willing to spend in anticipation of being able to recover the sums which he may agree to pay from the sale or use of the product upon which the work is done. Sooner or later the rate of wages is determined by conditions over which neither the employer nor the employed have any control. It is these forces which will be considered.

The purpose of this treatise is to determine the rate of wages expressed in terms of money. The distinction must be made between absolute wages and money wages. Absolute wages consist of the food, fuel, shelter, and savings if any, which are the true incentive to work. Money is merely the instrument wherewith absolute wages are obtained. Money serves to measure the work done, provided it be true money. If it be "mock money," as inconvertible paper money has been rightly called, it will serve to measure the work done, and in addition thereto the loss suffered by the workman, who is subject to the risk of the fluctuation in the purchasing power of the rate of his wages, which always ensues when inconvertible paper or mock money is forced into use in place of true money.

Before any intelligent consideration can be given to the determination of what makes the rate of wages, an absolute definition needs to be given to the word *money*. One of the great benefits which ensues from the study of economic questions is this necessity for the careful choice of words, for accurate definition, and for precision in the use of language as an instrument of thought or for the naming of things.

The Supreme Court of the United States has lately lent itself to the dangerous and fraudulent theory of *fiat* money. The Justices, save only one, have found in the sections of the Constitution which give Congress power to pass laws to enable the Executive "to coin money," or "to borrow money," reasons for yielding to Congress the power to coin paper and to make it lawful money. This decision is greatly to be regretted. It is replete with danger, and may yet cause much disaster to the people of the United States. Upon such a question as this, which is something more than a mere question of statute law, students and business men may rightly express an opinion, even if it is contrary to the dictum of the Court.

Great judges make precedents, and do not blindly follow them without consideration of the fundamental principles which must underlie all statutes, if justice is to be done by legal methods. When Mansfield declared that no slave could tread the soil of England, all precedents were against his decision. When Parsons ruled that none but free men could breathe the air of Massachusetts, he created a precedent, but he did not search for one. When Camden ruled that "general warrants" were inconsistent with English liberty, he went against the precedents of the courts for generations before. Yet, in making these decisions, these great judges brought the law of the land to the high level of the principle of human freedom, without regard to precedent. What had not been law until they so decided, became the accepted principle of law which no mere statute could afterward contravene. Had our Supreme Court but sought to give a true definition to the word *money*, they might have ruled that neither under the provisions of the Constitution for coining money nor for borrowing money could Congress or Court find authority for coining paper into money; or, in other words, for attempting to make something out of nothing. Had they given any consideration to the question, What is money? they would not have rendered a decision which, economically considered, is absurd, and by which they have substantially declared that the promise of a thing is the thing itself. Under this decision the people in this country have no rights which the Supreme Court is bound to sustain, if knaves or fools in these or other times pass Acts of Congress for stealing their wages or earnings from them by an issue of legal-tender *fiat* money. There were many ways open to the Court for sustaining the legality of a forced loan, without debasing the science of law or forcing an interpretation to the cases cited in the opinion, when these very cases, if rightly interpreted, are at variance with the decision in this case. It may be true that students differ, and that the definitions of economists are at variance with each other upon this question of what is money. The more reason for a Court of competent jurisdiction to give a definition consistent with right and justice, and to force all students or others who treat questions relating to money also to define the word in such a way that the substance cannot thereafter be confounded with the shadow—the thing for the promise of the thing carrying no obligation for the performance of the promise.

In this very question—the subject of this treatise—" What makes the rate of wages?" this recent decision of the Court must be ignored as if it had not been rendered, because it vitiates every form of statement which can be submitted. If the standard by which the rate of wages is established is liable to be changed at the instance of an accidental majority in any Congress, it ceases to be a standard. No scientific treatment of this or of any of the great economic questions now pending could be made consistently with such conditions; nor can any sound or permanent conclusions be reached consistently with this decision. So long as it stands, all acts of fiscal legislation will be of a purely empirical nature. If the opinion given by Justice Gray on behalf of the majority of the Court is to be accepted, that a national lie—a promise which implies no obligation to maintain it, is lawful; in other words, if a lie and a statute law can be consistent with each other, then truth, justice, history, and science alike reject and condemn the opinion by which such a conclusion has been reached. It would be out of place for an economist to venture to comment on the legal or technical grounds on which this opinion rests. Suffice it that in three trials the Court has been divided, and that there is as much weight of authority on one side as on the other, while outside the court it is difficult to find a lawyer of any high repute who sustains the present decision.

In this treatise it will therefore be assumed that no money is entitled to the name except standard coin, containing a fixed weight of precious metal. Between two kinds of coin there may be a distinction. One may be good money, the other may be bad money; witness our gold dollar and our base silver dollar of light weight ; but both kinds of coin are money, while no kind of paper promise can be money. Paper can only serve as a substitute for money. The standard by which we now work is the standard of gold coin ; but in the course of the treatise, many variations from this standard will have to be referred to, because during the period in which the country was subjected to the depreciated greenback currency, the rates of wages paid in terms of money served as no true guide to the absolute wages for which the work was done, as the purchasing power of this substitute for money varied with its own fluctuations, or in the ratio which it bore to the standard of gold coin. Among the minor evils of a vitiated currency is the uncertainty which is imparted to the statistical statements of the period in which it is used. Even a reduction of the currency prices of the war to a gold standard will only partially remedy this fault.

I am well aware that many economists of repute have adopted such a definition of the word money as to include any instrument of exchange which may serve the purpose. In so doing it may perhaps be held that they have given some foundation for the charge that political economy is not a science.

In the following treatise I have endeavored to prove the paramount importance of the question which serves as its title. Of what use would it be to treat the subject at all, or to attempt to analyze the forces which make the rate of wages, if there is no definite and established meaning to the word money in which the wages are rated ?

EDWARD ATKINSON.

BOSTON, MASS., U. S. A.,
July, 1884.

INTRODUCTION TO SECOND EDITION.

THE first edition of this treatise was pressed to completion rather hastily, with a hope that it might have some influence on the legislation of the present Congress in respect to Railroads and Silver Coinage.

A few errors have been pointed out by friendly critics, mainly owing to the slightly different results which are reached in reducing very large sums to rates of earnings per week or per day, without carrying out the decimals to such a point as would confuse the reader. None of these apparent errors affected the conclusions, and they have been corrected.

But it will be observed that only approximate accuracy can be claimed when the attempt is made to reduce the huge figures of estimated national production to the unit of what each person can enjoy each day.

Suffice it that even if the estimate of annual products be varied for possible error by ten per cent., or $1,000,000,000 (one thousand million dollars), the corresponding change in the share which each person may enjoy on the average each day would be only *five cents worth* more or less.

It has, perhaps, been a mistake, not to make a more complete separation of the theory of diminishing profits and increasing wages, from the statistics by which the theory is sustained; but as the work grew upon the writer's hands from what was intended to be a short essay, suitable for presentation at the meeting of the British Association for the Advancement of Science, into a treatise of many pages, the theory and its application became so interwoven that the writer himself could hardly have separated them in any greater measure.

EDWARD ATKINSON.

BROOKLINE, MASS., *Feb.* 19, 1885.

WHAT MAKES THE RATE OF WAGES?

THE phenomenal circulation, in England, of Henry George's book, entitled "Progress and Poverty," and the statement that it has already been translated into every civilized language although it made little impression in the United States, draws attention to the fact that all other questions have become relatively insignificant compared to the problems which relate to the distribution of wealth. The premises which Henry George assumes are without substantial foundation in fact and his conclusions are therefore without warrant. The *production* of what constitutes wealth or welfare is no longer at issue. Modern science and modern instrumentalities of production are adequate to produce what would suffice for a good subsistence for every man, woman, and child in any and all countries. The whole question at issue is the *distribution* of this substance after it has been produced. Production and distribution are but two phases of the same work.

Land, capital, and labor are the three factors in production, but even when these three factors are worked in the most hearty co-operation, the world is always within a year or less of starvation. The main question, therefore, is: How is the annual product distributed? because it is upon the distribution of the annual product that subsistence depends, rather than upon the ownership of land or of the products of labor which have been saved in a concrete form, and which have become capital. The capital or labor saved in a concrete form never exceeds in value the sum of two or three years production, even in the richest state or nation, and is more apt to be less than the product of a single year.

In the work of production and of distribution, by far the largest portion of the people of the so-called civilized world work for wages in one form or another,—that is to say, they are at any given time in the position of the employed rather than that of employers. They change from one class to the other, according to their relative abilities or opportunities. It follows of necessity that the paramount question—the one which is of prime importance to the vast majority of the people of civilized lands, is, *What makes the rate of wages?* because it is by means of the money which they receive from their employers as wages, that their share of each year's annual product is obtained and is measured. This being admitted, the practical question at once arises, are those who labor for wages receiving in each year a less and less proportion of the annual product, while capitalists are securing for themselves a larger share, or the reverse? Are the rich growing richer, while the poor become poorer? or, are nations themselves becoming poorer as a whole, rich and poor alike securing a decreasing share of a decreasing and, perhaps, insufficient product?

In treating this question, two definitions become necessary. What is *production?* It is not simply the primary process of bringing forth grain, timber, and metals in their crude form, from the field, the forest, or the mine; it is not simply carrying these products through the mill, the furnace, or the forge, into their secondary form, called manufactures; but the word must include all that is indicated by its etymology—*pro duco*—pro-duce-ing—leading forth and directing the forces of nature to the final use of, or consumption by, man. This covers *distribution*, as well as what is commonly called production. The word wages may, therefore, be defined so as to include all earnings of persons in the employment of others. The larger part of the work, in many directions, being done by the piece, the wage is an uncertain quantity, varying with the skill and capacity of the laborer. In this treatise the word *wages* will stand for the sum of money which is earned by factory operatives, farm laborers, machinists, mechanics, railroad employees, laborers, clerks, salesmen; in fact, by each and every class of those who are employed by others in what is commonly called production or distribution: those who agree in advance to work for a fixed payment, either by the piece or by the day, month, or year.

The true wage which the workman seeks is the food, fuel, shelter, and other means of subsistence with which the sum of his wages will supply him. If we look to the derivation of the word itself, his wage is the measure of the expectation of subsistence, against which his labor is staked, wagered, or hazarded. It is not customary to include the salaries of the clerical or administrative force, nor the payments which are made for purely mental work under this term, although they are of the same nature. For the purpose in hand, we will limit the application of the word *wages* to the sum of money earned by persons who engage in the actual work of producing or distributing material substances; who either work with their hands or direct machinery to these ends; who are in the employment of other persons upon terms stipulated in advance and who are subject to be discharged with or without notice, as the case may be, at the will of the employer. In this category will be found by far the largest portion of the people of this country who are old enough to become wholly or in part self-supporting.

This great class consists in very large measure of persons who depend almost wholly upon their daily work for their daily bread,—whose accumulations are small,—slowly and painfully made or saved, and sufficient only to relieve them from the necessity of work for the last few years of old age, if perchance adequate for that without the aid of their children. The welfare of the vast majority of the people of this country, and of every other country, therefore, mainly depends upon the adequacy of the rate of their wages and upon the purchasing power of the money in which their wages are paid. It follows that there can be no more important social question than the wage question,—none in which error will be more fatal.

If, under the existing conditions of employer and employed,—of capitalist and laborer,—of wage-payer and wage-receiver,—in other words, if by way of competition the rich only grow richer because the poor grow poorer;—if greater progress under present laws and customs is only consistent with greater poverty;—if the profits of capital can only be increased by diminishing the wages of labor;—if "wealth accumulates only when men decay," —then socialism may be justified, even nihilism may be right; the capitalist may be the enemy of the laborer. If such is the truth, Henry George only goes half way in his remedy, when he merely proposes to nationalize or confiscate land. The remedy for these great apparent wrongs may, in such event, be found only in dynamite and the dagger. If even the change in institutions or in the title to land which can be secured by legislation is insufficient, then dynamite and the dagger may be the only adequate remedy, as Wendell Phillips hinted, but even he dared not say so, in his Phi Beta Kappa oration. The very existence of modern society is the major issue which is bound up in the simple and apparently minor question, "*What makes the rate of wages?*" Compared with this all problems relating to the collection of revenue, the function of banks, the hours of labor, etc., sink into relative insignificance. If the fundamental question is, What makes the rate of wages?—these minor questions are merely the froth and turmoil upon the surface, which manifest to the eye and ear the great undercurrent which may rend modern society in twain.

What are the facts? Upon the continent of Europe, ancient forms of society, customs, laws, and institutions of many kinds, from which we in this country are substantially free, are being actually rent and destroyed, and the whole socialistic tendency of legislation at this time, in Great Britain, France, Germany, and elsewhere, is but an attempt to solve the apparently simple question, What makes the rate of wages, or of the earnings of those who depend upon their daily work to meet their daily wants? By socialistic tendency is meant such acts of legislation as the Land Acts relating to Ireland lately passed by the Parliament of Great Britain; the acts for compulsory life or annuity insurance which have been proposed by Bismarck; the attempts which have been made in France to own and control the whole railway system and to maintain national workshops; and many other measures of like kind which have been either proposed or attempted in different parts of Europe. The issue is made more difficult by the existence of conditions in Europe to which we have nothing analogous. The question there is not only: What makes the rate of the wages of the factory operative, the mechanic, or the artisan? but, What makes the rate of earnings of the Irish cottier, or the rack-rented farmer, or of the English tenant farmer working leased land; or of the French or German peasant confined to allotments which have been mainly established by the compulsory division of land on the Continent, and which have become so small by frequent subdivision that modern agricultural machinery cannot be applied to them in any great measure; on which the crops are therefore made by the exertion of the maximum amount of manual labor with the minimum of product per man? An example may be here cited of the vast difference, in different places, in the productive efficiency of one man, working one year. I cannot give the exact measure per man in bushels of grain or barrels of flour of foreign agriculture, but the German or French peasant makes but a very small crop, who, with arduous toil with the spade and hoe, plants a little strip of grain, harvesting it with the sickle, and thrashing it with the flail; every one can conceive how small a quantity of grain must be the product under these conditions, yet these are the conditions under which a considerable, if not the larger portion, of the grain crops of Europe are made.

On the other hand, let us consider an extreme example of the application of capital to great areas of land in this country. By division of labor and by the application of machinery upon the great farms of Dakota, such enormous abundance is secured that when we convert bushels of grain to the equivalent of one man's work, working 300 days in one year, we find that in an average year, on land producing twenty bushels of wheat to the acre, 5,500 to 5,600 bushels of wheat are made for each man's work. Retaining enough for seed, this quantity suffices to make 1,000 barrels of flour. It can be carried through the flour mill and put into barrels, including the labor of making the barrel, at the equivalent of one other man's labor for one year; and at the ratio of the work done to each man employed upon the New York Central Railroad, the 4,500 bushels of wheat can be moved from far Dakota to a flour mill in Minnesota, and thence the 1,000 barrels of flour can be moved to the city of New York, and all the machinery of the farm, the mill, and the railroad can also be kept in repair at the equivalent of the labor of two more men; so that the modern miracle is, that 1,000 barrels of flour, the annual ration of 1,000 people, can be placed in the city of New York, from a point 1,700 to 2,000 miles distant, with the exertion of the human labor equivalent to that of only four men, working one year in producing, milling, and moving the wheat. It can there be baked and distributed by the work of three more persons; so that seven persons serve one thousand with bread.

Before we proceed further in the consideration of this and other related facts, let me say that there appears to be an almost unacknowledged belief, even among well-read students, that the so-called principle which Malthus first propounded is true; or at least that it contains such an element of malignant truth, if one may use such an expression, that it is unpleasant to face it, lest one's faith in the Power that makes for righteousness should be disturbed. If the dogma of Malthus is true, that population tends to increase faster than the means of subsistence, there is no escape from the conclusion that all our efforts at progress, so-called, are worse than useless; for instance, when we attempt to save the life of children by the improved sewerage of our cities; when we provide pure water and better dwellings for the poor, when we teach sanitary science to enable each and every member of the community to attain present better conditions of comfort and welfare and a longer life, we are merely building up our present prosperity in order that the adversity of a future day may affect a greater number of people. If population increases faster than the means of subsistence, the rate of wages must always tend to become a less and less proportion of a decreasing product and their purchasing power must at last become so low as not to assure even the necessary subsistence; because there would not be substance enough to sustain life to be purchased by any wages which could be paid. In such a view of life all our humanitarian efforts are criminal if successful, because they cause a more rapid increase of population and only hasten the evil day when, in spite of every effort or of any measure of intelligence, our mother Earth will fail to provide for the wants of her children. They must then slay each other or die in myriads by famine and pestilence, in order that only the fittest may survive. Even then, when those only have survived for whom there is enough for the moment, the evil cycle would begin once more and so go on forever. It is upon the seeming truth which is contained in this abhorrent and atheistic dogma that many false theories have been presented, many bad acts of legislation have been justified, and that it has become a widespread conviction that there is a war, or constant struggle and antagonism between capital and labor,—between rich and poor. It seems to be the conviction of great masses of people that with ever increasing wealth there is and must be ever increasing poverty, and this formula is working in special places in the most active and pernicious manner at the present time. Again we may ask, what are the signs of the times? Russia struggling with nihilism; Vienna under martial law, for fear of socialism; Germany and Austria dreading what may come when Bismarck dies; the commune of Paris kept down only by fear and bayonets; even England, gravely disturbed by a single book which attacks her land system, is coping with Irish destitution by acts of Parliament which are but socialism disguised and which would be overruled, if enacted by the Congress of the United States, the moment they were presented to the Supreme Court. These dangers to the body politic are signs that the struggle for life has indeed become urgent among great masses of people in special and limited places. They indicate that even in the present day the horrors of the Reign of Terror might be repeated; that want is lawless; that hunger and destitution will incite to violence in any land; and they also prove that the more the attempt is made to suppress these dangers by force of arms, the greater the danger will become. It would be as dangerous to disband the armies of Europe as it is impossible to sustain them, because the habit of government by force cannot be overcome except after many years. Yet, as I have said, in the world there is always enough. Production is ample to give good subsistence to every man, woman and child, especially in the civilized world, and the mechanism of distribution is also fairly adequate. The whole question is one of the method of distribution of each year's product, and inasmuch as this distribution is mainly effected by way of the payment of wages, the paramount question is again presented:

WHAT MAKES THE RATE OF WAGES?

If we glance again at the condition of the nations which have been named, we cannot help observing, for instance, that Germany is poor in fact; the soil of large portions of her territory will barely sustain the people who dwell thereon, and although there has as yet been no absolute famine, the people of many parts of Germany are always on the very edge of want. We must therefore explain to ourselves the conditions of danger to which the best instructed people of Europe have been brought, by the consideration of other matters. The people of Germany must be subsisted either upon what her own soil will produce, or upon the food for which her own manufactures will exchange. Her own annual product, at its exchangeable value in money, must be the source of her own profits, wages, and taxes. When we utter the last word, may we not touch one secret of her poverty? There are money taxes and also blood taxes. One man in every twenty in Germany is a soldier in camp or barracks, and one other man in every other twenty must be employed in sustaining the idle soldier, while every man wastes a considerable part of his life in preparation for this destructive art and is liable to be called away from productive work at a moment's notice. Under such conditions, before either profits or wages can be paid to those who do the work, at least ten per cent. must be assigned to the wasteful and destructive although generally passive war which is the condition in which all the nations of Europe now exist.

How is this army maintained? There is room enough elsewhere, and to spare, for Germany to relieve herself of the population which cannot live upon her soil, except on the edge of starvation; there is room enough even in our own land and here they would be welcome. But every German boy who reaches the age of eighteen is enrolled for service in the army at a future day, and if he dares leave the country after he is enrolled, he expatriates himself, renders any property which may be devised to him liable to confiscation, and can never return, even though he may have become an American citizen, except at the risk of being treated as a deserter, and forced to render his three year's service in camp or barracks. Under such conditions as these it follows that neither the poverty of Germany, France, Austria, Italy, nor any other country, can be attributed to any real antagonism between labor and capital, but must be attributed in part to the poverty of the soil, in part to artificial systems in the division of the land which are enforced by statute and in part to privileges and to the burdens of standing armies of which we have no counterpart. These dangers to the body politic are but signs that the struggle for life has indeed become urgent among masses of people who number too many for the limited area in which they are, but where they are kept by force, the natural law of distribution by which they might spread themselves over the earth being obstructed. Much of this is done under the pretext that the right to property can only be permanently sustained by force, while the rights of man are denied.

We may also observe that almost all modern dangers of war are dangers connected with the distribution of wealth, or from national jealousy in respect to commerce which is but another name for the distribution of the annual product of the world. This jealousy is mainly caused by the continued prevalence of the false idea that in international commerce what one nation gains another loses. Hence we find nations endeavoring to establish and maintain colonies, in order to control their commerce, at a cost to themselves of more than the whole commerce is worth.

No one fights to-day for a religious dogma, unless it be an Arab or a Sepoy. None are armed merely to maintain a dynasty. It is the Chancellor rather than the Emperor on whose fate the Empire of Germany may depend. The question as to who shall control the Suez Canal endangers the peace of Europe, yet this canal is but a spout through which Europe exchanges clothing for food; it is a mere instrumentality of distribution. All modern questions of any importance relate to the means of subsistence; the distribution of the means of subsistence is finally brought about by the payment of wages. The first question which England has met in endeavoring to promote good government in Egypt, is the debt incurred by a despotic power but imposed on the people who were oppressed. Whether the repudiation of such debts is not the first condition precedent to the common welfare of those upon whom the debt has been imposed without their consent, is one of the many questions about to be forced to an issue in other countries than Egypt. If one half the product of Egypt is absorbed by the debt, will the other half suffice even for subsistence? Can the sum of wages be more than what is left of her own product? Must not the annual product of each country be the source of its own wages?

As I have said, when we attempt to solve this question, we find that there need be no fear of want because there is not enough for all. Enough there is, and to spare. The only question is, Where is it? Distribution is limited or restricted in part only by want of proper mechanism, *i. e.*, by the lack of railways, the lack of ships, and the like; in part by legal obstruction, in part by national jealousies, but yet more by obstacles to free exchange, even where the mechanism suffices. I do not limit the term *free exchange* to the narrow question which is at issue between the advocates of free trade and protection; that is a minor question. I mean the obstacles to free exchange which are mainly caused by that ignorance and incapacity which stands in the way of mutual service, even among the people of the same country. The farmer of our own land may have his barns running over with the abundance of his product, and may desire a hundred things for which he would be willing to exchange; but if, on the other hand, those who desire to share his abundance are ignorant, incapable, or vicious, who cannot or will not work upon the things the farmer wants, there can be no mutual service: they may starve while his crops decay. It is mainly the imperfect or restricted distribution of what there is ready for use, which is caused by the ignorance or incapacity of those who need it, that creates want in the midst of plenty, not only in Europe, but in the heart of the great cities of our own land. We waste enough in this country to support all our poor in luxury; yet were we to give this excess to them in mere charity, what we waste, thus consumed, would forever convert the poor into paupers. Charity or alms-giving cannot remove pauperism; it may only increase it. The *common* laborer, so called, is the one who suffers most in times of depression; and he usually is and remains a *common* laborer merely because neither his hand nor his head have been trained together so as to enable him to do work requiring skill, which kind of work is everywhere and at all times waiting to be done, and by doing which he might become entitled to a share of existing abundance. We are attempting, in this country, to cope with these problems by legislative methods. In Europe the attempt is made both by legislative methods and by force combined. Neither method can permanently succeed. Neither wealth, welfare, nor common subsistence can be permanently imposed from above, or instituted from without. Neither masses of men nor individual men can be permanently helped who cannot or will not help themselves. The final remedy for these wrongs can only come by the development of individual manhood from within. Individual intelligence and integrity, sustained by public justice, constitute the sole condition under which permanent prosperity can become the rule among men. Then life and liberty will be the only common factors, making for the welfare of each and all. It may be a far-off day, which none of us living may live to see, when this shall be accomplished; but the potential agency in promoting this end is *the advancement of science.*

With the chemical or physiological question which underlies the abhorrent dogma of Malthus, I may not attempt to deal. Subsistence is but a conversion of forces—a chemical process; whether or not the proportion of force or energy which constitutes material life, and which takes the form of the body in which man lives awhile on this earth, may find a limit without recourse to war, pestilence, or famine to check its undue development, is not yet a practical question. When it arises, it may be time enough to meet it, in some far away period.

The absurdity of the attempt, as yet, to measure the power of subsistence and to declare it to be limited can be demonstrated in two or three simple ways suitable to the use of a statistician like myself: First, no man yet knows the productive capacity of a single acre of land anywhere in respect to food. Second, the whole existing population of the globe, estimated at 1,400,000,000 persons, could find comfortable standing-room within the limits of a field ten miles square. In a field twenty miles square they could all be seated, and by the use of telephones in sufficient number they could all be addressed by a single speaker. Third, the average crop of wheat in the United States and Canada would give one person in every twenty of the population of the globe a barrel of flour in each year, with enough to spare for seed; the land capable of producing wheat is not occupied to any thing like one twentieth of its extent. We can raise grain enough on a small part of the territory of the United States to feed the world. The great American desert has gradually disappeared. The "bad lands" of Montana prove to be the best grazing ground of the Northwest, and in the heart of the Eastern States the mountain section of the South waits for a population equal to that of Great Britain, who can there find potentialities in agriculture or in mining equal to those of any similar area on this or any other continent. As yet, therefore, the doctrine of Malthus has found only a limited application, where some local or temporary congestion of human force has gathered. As I have said, in the world there is somewhere and always enough. The only question is, Where is it? When found, the next question arises, How to get it?

The first method which obtained in the world, was to grab it—the age of force. The second method was to give it—the era of conqueror and conquered, of master and slave, of lord and vassal, of giver and taker, not of employer and wage-earner. The third method is to exchange for it. Under this third method commerce has arisen, men have become sorted as capitalists and laborers, as employers and employed, as wage-payers and wage-receivers; service for service is the common rule of life; the exchange of product for product is the practice of commerce. All States have, or may become interdependent, and then "the ships that pass between this land and that will be like the shuttle of the loom, weaving the web of concord among the nations." And again we meet the apparently simple question, What makes the rate of wages by which the greater part of these services are measured and under which the greater part of the distribution is effected?

I have had but little time for the reading of books or the consideration of theories of wages; but I believe we must pass from the English orthodox system of political economy to France, in order to find the first true statement of the relations of the wage-receiver and the wage-payer, of employer and employed, of laborer and capitalist, or of labor and capital. Many years ago a single phrase in Bastiat's "Harmonies of Political Economy" became engraved upon my mind, and by its application I have been enabled to observe the phenomena of wages in the course of my business life with much clearer insight. It is this: "*In proportion to the increase of capital, the absolute share of the total product falling to the capitalist is augmented, but his relative share is diminished; while on the contrary, the share of the laborer is increased both absolutely and relatively.*"

Among English writers, Thornton exposed the fallacy of the old wage-fund theory, the theory that all wages are paid out of a fund of capital previously accumulated and will be high or low as the ratio of that fund may be great or small, in proportion to the number of persons employed. Professor Cairnes propounded the true theory of wages in one of his latest books, in terms so nearly identical with some of those which the writer had used in this treatise, that the writer would have suspected himself of unconscious plagiarism had he not found his own records antedating the published works of Professor Cairnes on this subject. In this country, Professor Francis A. Walker has presented the true theory of wages in the most effective manner and has probably done more than any other writer to clear the subject of obscurity. It has been a matter of great satisfaction to me, that my practical observations are so fully consistent with the theories of these authors. Giving due credit to all these writers, my own conclusions have been based almost wholly upon facts and deductions from business experience rather than from books, although my attention was first attracted and a direction was given to my observations by the paragraph which I have quoted from Bastiat.

The two forces which are engaged in the production of the substances which constitute food, fuel, means of shelter, or the materials which may be converted into additional capital, are of course, labor and capital. Land itself is but an instrument, being useless and valueless unless labor and capital are employed upon it. By the co-operation of these two forces, an annual product is made. The true function of capital is that of a force put to use in order to increase production, rather than a substance to be immediately divided and consumed.

Fixed capital, so called, although the name is hardly a suitable one, may be likened to the foundation, boiler and engine, and quick capital to the fuel with which the boiler is supplied: the one is very slowly, the other very quickly consumed, yet neither works directly to the subsistence of men, but indirectly both work to the vast increase of the actual substances with which men are fed, clothed, and sheltered; these substances constitute the annual product which is divided among them. The term *annual* fits the case, because the year represents the course of the four seasons and the succession of crops. A small part of each year's annual product, commonly called " quick " or " active capital," must be carried over to start the next year's work upon, as a small part of last year's product had been brought over to start this year's work upon; one proportion balancing the other. The fixed capital seldom exceeds in value two year's production. It therefore follows that all profits, all wages, all taxes, in fact all consumption whereby existence is maintained, must be substantially drawn from each year's product; it is therefore in the division of these substances produced within the year, that true profits and real wages are to be found. But, in order that this product may be distributed and consumed, since no man lives, economically speaking, for himself alone, the various products of the year must all be exchanged by purchase and sale, and therefore must all be measured in and reduced to terms of money,—except that part of the annual product which is consumed upon the farm by the farmer and his family without being sold. With this exception, it therefore follows, that substantially the whole product of each year must be converted into terms of money. I think it escapes common observation, that in all departments of industry, except agriculture, few men now produce any thing which they use themselves; and even in farmers' families, domestic consumption is now limited to a small part of the farm product, all else is procured by exchange; all men are interdependent. The sum of money represented by this conversion is and must be vastly greater than the sum of real or actual money which is used as the instrument of exchange, hence the necessity for true money. The greenback fallacy can only deceive those who fail to comprehend the function of money. Inconvertible paper money is a fraud, and the burthen of proof rests upon its advocates to justify the honesty of their intentions by the weakness of their intellects. In this process of conversion into terms of money by way of purchase and sale, a part of the value of the annual product is sorted on the one side as profit, rent, interest, or by whatever name the share of the owner of capital may be designated; and, on the other side, another and vastly greater part constitutes the share of those who do the work, and is named wages. In the subdivision of this latter share into individual parts, the rate of each persons wage is established in terms of money.

It would not be consistent with the general purpose of this treatise to attempt at this point to give precise details in respect to the value of the annual product of a normal year in money. The general conclusion at which I have arrived is, that in the year 1880, the census year, when the population of the United States numbered a little over 50,000,000, the annual product had a value of nearly or quite $10,000,000,000 at the points of final consumption, including, at market prices, that portion which was consumed upon the farm but which was never sold. Omitting that consumed upon the farm, it was about $9,000,000,000. What portion of this product constitutes the average share of the capitalist at the present time cannot be substantially proved. In a normal year, under normal conditions I am of the profound conviction that not exceeding ten per cent. can be set aside as either rent, interest, profit, or savings; and that nine tenths constitutes the share of the laborer, which, by subdivision, becomes expressed in terms of personal wages.

During recent years, the increased efficiency of the railway service, and the consequent elimination of two thirds of the cost of distributing commodities in bulk, has undoubtedly augmented for a time the amount falling to the capitalist, but without in any measure reducing the amount previously falling to the laborer; on the contrary, greatly promoting the laborer's interest as well as that of the capitalist.

The great fortunes of the railway magnates (aside from one or two conspicuous and notorious thieves who have stolen franchises and defrauded their stockholders) have consisted of but a small portion of what they have saved to the community. The main work of railway capitalists has been to reduce the cost of distribution; their true function ought not to be prejudiced by the fact that a judge of one of the courts of a neighboring State was impeached and disqualified from holding any office of trust or honor for "corrupt practices" with a notorious railway official. The corrupt judge is dead—the corruptor of the judge still lives a base and dishonored life, probably continuing to exist physically because he is mentally and morally incapable of conceiving the turpitude of his existence or of feeling the loathing and contempt of the community. But even the railways which he has constructed will continue to serve some useful purpose after the corruption which he has engendered has been buried with him in a nameless grave.

In treating this question of the rate of wages, it must constantly be kept in mind that money is but the instrument of exchange, that real wages are what the money will buy, and there cannot be more real wages than the whole product, less the share of capital. If then, we can even approximate the value of the product and divide by the known number of persons employed, we then approximate the annual measure or average rate of wages in terms of money.

At the risk of repetition this point must be further considered, as it is the key to this treatise.

The population of the United States, in the census year, consisted of a little over fifty million persons, or about ten million families of five each. Substantially one in every three was engaged in some kind of gainful occupation. Agriculture was and is the leading occupation. Upon small farms, a large portion of the produce is consumed by the farmer, his family, and his laborers. Upon large farms, the greater part of the produce is sold. In the families of country mechanics, much productive work is done which in cities is procured by purchase. We can only approximate in a general way the value of the domestic consumption. If one tenth of the consumption of the country is of the nature of purely domestic production and consumption, which is never converted into terms of money by purchase and sale, the total sum which would represent such domestic consumption would be $20 to each person, $100 to each family, or $1,000,000,000 total value. Of this the census enumerator would find no trace in the figures of commerce. This is a large estimate, undoubtedly, of the domestic consumption of articles which might be or might have been procured by purchase, but which were in fact produced and consumed without purchase or sale. The remainder of the annual product, at whatever sum of money it may be finally valued when sold for the last time and distributed for final consumption, constitutes the value of the product converted into terms of money, from which sum all money profits, all money wages, and all money taxes must be derived. There can be no other source. Each bargain for a sale or a purchase is and must be made in terms of money. The manufacturer, the merchant, and the shopkeeper take their toll of profit in money, not in kind. The assessor levies a tax payable in money. When this tax is levied upon a producer or a distributor, it is charged to the cost of the business, and is thus distributed among those who buy the goods for consumption. The laborer receives his wages in money, seldom in kind, except the farm laborer; he then converts his money into his share of the annual product by the consumption of which he sustains life. The total sum of money which represents the value of all that is produced, at its point of final consumption, is and must be the final measure of that part of the annual product which is bought and sold. Therefore, all profits, wages, and taxes constitute a portion of this lump sum; in order to ascertain what the rate of profit, the rate of taxation, or the rate of wages may be, we must ascertain what this lump sum is, and how it is divided. On the other hand, by ascertaining what the total sum of taxes, the sum of all wages, and the sum of all profits may be, we can again approximate the total value of the annual product. No absolute results can be reached by either method, but approximate results can be fairly set off, one against the other. This is what the writer has endeavored to do.

The principle which I have attempted to sustain in this

treatise may be considered without any regard to its application to the existing figures of the present date. I have given these figures, however, in the way of an illustration. They will be more fully treated in appendix I.

The principle might be stated in algebraic symbols. For instance, given the question, "What is the value of the annual product of the year 1884?" It would consist of the following elements: First, the wear or consumption of fixed capital previously accumulated; the proportion of the quick capital or product of the year 1883 brought over to and consumed in the year 1884, in order to begin work. Let these two elements be called a. To them would be added the actual product of the year. Let this be called b. From this product a certain proportion would be carried over, to begin the work of the year 1885. Let this be called c. The formula could then be stated in the following terms: $a + b - c = x$, the annual product which is subject to subdivision and to consumption.

Let profits be called d, sum of all wages e, persons engaged in gainful occupation for a given rate of wages, f, and the average rate of wages i. The complete formula would then be as follows:

$$a + b - c = x$$
$$x - d = e \div f = i$$

If i be the average of all there is, one wage earner will earn less, another more, according to relative capacity and opportunity, and by competition each with the other: but these earnings, differing each with the other, will be absolutely within the limit of i; while i itself will annually stand for an increasing share of an increasing product, if my premises are sustained.

In a computation of what makes the total accumulated wealth of the United States, which was made by the Census

Department, one half the value of the product of mines, oil wells, and the like, was taken as being on hand at a given time, constituting a part of the accumulated wealth, together with three fourths of the annual product of agriculture and manufacturing. Working from these data, it appears that the census estimate of the value of the annual product of the United States for the census year was from $8,200,000,000 to $8,500,000,000, not including domestic consumption. There appears to be no actual computation of the value of the annual product in the census, but the figures used in the computation of wealth yield these approximate results. The writer had reached his own conclusions by very different methods from those used by the Census Department, and had satisfied himself that if there be added to that part of the annual product which is sold, and which is, therefore, reduced to terms of price in money in the markets of the world, the domestic consumption upon farms and in families, the total value of the annual product would not exceed $10,000,000,000 in the census year, at the retail prices for final consumption. If the census estimate be divided by the population of substantially 50,000,000 people, we reach $160 to $170 per year as the sum representing the average annual product for each person, or a fraction less than forty-four to forty-seven cents per day for 365 days. That is to say, when the products or services of each person were brought into competition in the markets of the world, the money value of the entire commercial product in the census year was measured by the average sum of forty-four to forty-seven cents' worth to each person. My own computation gives a little under $200 to each person, including the domestic consumption of farmers, or a little under fifty-five cents' worth per day. That is to say, the average product of each person may be estimated by any one who will go into the market, hire shelter, procure food and clothing, and save something out of what fifty-five cents a day will pay for for each member of a family. If no more is produced, no more can be had. What there is may be bought and sold ten times over; it only wastes a little each time; it does not increase. Paper may be substituted for true money, and the rate of paper wages may be apparently doubled, but then it will take $1.10 in paper to buy what fifty-five cents gold now buys. There cannot be any more shelter, food, fuel and clothing sold than there is produced, and the value in money of all that there is produced is the final measure of all profits and wages. The subdivision of all there is produced, therefore, *makes the rates of both profits and wages.*

If, again, we call $1,000,000,000 the domestic consumption, and value the salable portion at $9,000,000,000, and then divide by the whole number of persons in productive work (excepting soldiers and minor Government employés), to wit, 17,300,000, we reach an average of $520 as the annual measure of the productive services of each person thus engaged in useful work, each one at work sustaining two others. This computation may be proved to be substantially correct by a comparison with the actual wages or earnings of all classes, which were treated separately in the census, giving due consideration and applying judgment to the relative value of the work done. (See appendix I. for exact comparison.)

It may, therefore, be assumed that the average value of the gross product of each person who was engaged in any lucrative or productive employment in the United States in 1880, can be fairly established in the census year at a sum closely approximating $520. If such is the measure in money of all that was produced, then all wages, profits, taxes, and all savings or additions to capital must have been derived from such a sum. There can be no other source for either, unless the country incurred a foreign debt, which it did not in any great measure. It paid more debt in the census year than it incurred.

If such is the gross sum, let us see what the net sum free from taxes, may have been. In the same census, the gross sum of all National, State, county, and municipal taxation, was computed in round figures at over $700,000,000, or over $40 *per capita* of all persons engaged in gainful occupations. If we apply this rate to the average share of the product which fell to each person who was occupied in gainful occupation, we reach the following result: Gross product, $520; taxation a little under 8 per cent., $40.00; net share of the annual product, free of taxes, valued at $480. Now it will be apparent if only one in 2.90 persons is employed in gainful or productive occupations, then 2.90 persons must be subsisted upon what $480 per year, or $1.32 per day, will purchase, or 45½ cts. worth to each person; if it be considered also that from this sum must be set aside profits or additions to capital which take precedence of wages or earnings, then it will at once appear that by far the larger part of each year's product must be consumed; that is to say it must enter into the cost of production. In point of fact each year's work barely suffices for each year's wants and but little can be saved or added to capital because it is evident at a moment's consideration that not much can be saved out of what 45 cents will buy for each person each day. There is no absolute method of determining the exact proportion of the annual product which can be set aside as profit or addition to capital, nor of ascertaining that part which constitutes the actual wages or earnings. All that can be said is this: If 10 per cent. of the gross product can be set aside in a normal year, for the maintenance or increase of capital, that is to say, $48.00, out of each person's net share of the whole, then the average rate of wages or earnings of all the people of this country engaged in gainful occupation, is at the rate of $432.00 per annum, $1.19 per day or $1.44 per working day. This result, again, fairly approximates to the disclosure of the census, if it be compared with the specific ascertained earnings of persons engaged in special branches of industry. If any thing, it is a large estimate rather than a small one.[1]

If the foregoing premises be admitted, it follows of necessity that so far as those who work for wages are concerned, the relative or proportionate rate which each one or each class may receive cannot be in any very large measure affected by the sum which is set aside as profit or increase of capital, but must be mainly affected by the competition of laborer with laborer and will be finally determined by the relative efficiency of each person within the limit of the average proportion which his class receives out of the annual product. That is to say, the relative condition of each class of laborers must be determined by the variation from a standard or average which is determined by the quantity and price of the aggregate product of that class, *i. e.*, in that special branch of industry. The general rate of wages can therefore only be raised by an increase of product coupled with a wider market commensurate with such increase, so that the price may be maintained. Absolute wages may be increased although the rate in money may not, by an increase in the product, accompanied by a decrease in the price, so that the same or a less rate of wages may buy more commodities. The gross product may be increased by two methods only; first, by the intel-

---

[1] See table of earnings or wages in appendix.

ligent use of the increase of capital; and second, by the more intelligent co-operation of labor with capital. Contention or antagonism can only result in diminished rates both of profits and of wages. Prices and rates of wages can only be maintained by enlarging the market as labor becomes more effective and a greater quantity of things is produced by a decreasing number of persons. When a greater quantity of any given product is made by an improvement in machinery or a new invention, and men who have before been employed in that art are no longer wanted—then a wider market must be found for products which remain within their capacity to produce. Hence, those nations which apply machinery in greatest measure, and thus increase the quantity of their product while diminishing the cost as well as the number of persons employed, possess the greatest power of competition in supplying other nations in which all the arts are mainly handicrafts. For instance, England and the United States compete with each other in supplying China with a portion of the cotton fabrics needed by the Chinese (supplying perhaps ten per cent. of the cotton fabrics which are consumed in China) in exchange for tea, silk, etc., etc. The cultivation and preparation of tea and silk being of necessity handicrafts, this exchange would occur even if no climatic condition entered into the case. The exchange of fabrics made by machinery for tea and silk, yielding each nation what it needs with the least effort, although the quantity of labor varies greatly.

It therefore follows that the power to control commerce with the non-machine using races, who constitute more than three fourths of the population of the globe, rests with that nation which applies machinery most effectively to the greatest natural resources, and whose product is least diverted from being applied to profits and wages by destructive taxation, such as the support of a great standing army or costly navy.

The invention of machinery creates commerce. If we revert to the former conditions of life in the different sections of the United States, may we not find an explanation of the vast increase in the domestic commerce of the country, in the greater *interdependence* of each section of the country upon each other section, as well as in the greater interdependence of individuals upon each other. Exchanges of product for product have widened and increased, perhaps in greater measure than the aggregate product itself. If we recall the conditions of life of the New England farmers and artisans in the early part of the century, a very small money income sufficed them, because they lived mainly upon what they produced themselves, and because many of their exchanges were made without the intervention of any money. They *swopped* or bartered services in the erection of their dwellings and in harvesting; they raised, spun, and wove their own wool; they packed their own pork; they raised their own corn and paid for grinding it by a toll in kind; they cut their own fuel. These primitive conditions can even now be observed in the mountain sections of the Southern States. But even under such conditions, the consumption of food and fuel of each person may not have varied greatly in quantity or weight from that of the present time. It differed greatly in kind and in quality, and also in the method by which it was attained; but the quantity of food in ounces, which is the final standard, cannot greatly vary in one period as compared to another. We waste a great deal more now than we did in those early days, but our actual consumption of food per person cannot have increased in any very large measure. In the primitive days, under these primitive methods, the labor was so arduous and the hours of work were so continuous that only the strongest survived. The figures representing commerce were very small and when wages were paid at all, they were at very low rates for long hours of merely manual labor. Under the modern method of extreme subdivision, and the application of adequate machinery, *i. e.*, capital, the labor is less toilsome, the hours of work are shorter, the weakest can find something to do, each serves the other, and in the process of manifold exchanges, the figures representing commerce rise to almost incomprehensible millions; yet the actual quantity consumed, as I have said before, may not have varied in any great measure, so far as food and fuel are concerned. So far as clothing is concerned, production and consumption have increased enormously.

The end of all this vast system of exchange is, however, that, in one way or another, each person may secure about three pounds of food per day, a few yards of cotton or woollen cloth each year, two or three tons of coal or five or six cords of wood a year, and a given number of cubic feet of space, sheltered by a roof. They needed as much per person of the absolute necessaries of life fifty or a hundred years since as they do now, but they obtained them only by working twice or thrice as hard. They were more independent, less interdependent. There was far less capital, and much more arduous and excessive labor. The conditions of life were more equal, but it was the equality of sordid, continuous, excessive manual labor, aided neither by the factory nor by the railroad; neither by the more modern inventions of the masters of science, nor by the administrative and organizing power of the great capitalists, without whose potential work all modern progress would have been substantially impossible. The fortunes which those great directors of industry have made for themselves bear but the proportion of a small fraction to the labor which they have saved their fellow-men.

I will repeat again what I have said before: the late Cornelius Vanderbilt may be taken as an example of a communist in a true sense. He was the greatest communist of his age. He consolidated and perfected the railroad service in such a way that a year's supply of meat and bread can be moved one thousand miles, from the western prairies to the eastern workshops, at the measure of cost of a single day's wages of a mechanic or artisan in Massachusetts—that is to say, if the mechanic or artisan of the East will give up one holiday in a year, he removes one thousand miles of distance between himself and the main source of his supply of necessary food.[1]

---

[1] I have cited the late Cornelius Vanderbilt as the great communist of his age for the reason that he may be said to have first invented the consolidation of a through line of railway from the prairies of the West to the markets of the East, with a consequent reduction in the cost of bread and meat to the dense population of the Atlantic seaboard. By this consolidation and effective service, one thousand miles of distance have been substantially overcome at such a small cost as to have rendered the choice of position, at any point within that range, a matter of so little moment in respect to the supply of Western food as to be practically out of consideration. For instance, the value of the product of five hundred operatives in a coarse cotton factory in Massachusetts is over one million dollars—all the western flour and meat which these operatives need in a year can be moved from Chicago to Lowell at a cost of $600, and sometimes for less.

It is sometimes urged that such great fortunes as that of Vanderbilt and a few others are against the public interest, and that some method ought to be devised for limiting their accumulation. This ungrounded prejudice has mainly arisen from the jealousy rightly caused by the great fortunes which were accumulated by expert gamblers under the malignant system of the greenback or legal-tender paper money before these notes had been made redeemable in gold coin.

It is very true that the most of the fortunes which were made out of the fluctuations of the currency were speedily lost, but the foundations of a portion of the most conspicuous existing fortunes were laid under these bad conditions.

It is hoped, and may be believed, that advocates of paper money will

Having attempted to estimate the main factors which determine the general or average rate of wages at a given time, we may now consider the subdivision or the forces which affect the subdivision of the true wages fund. Why is the average rate of wages in a given occupation two dollars a day in one place, and one dollar a day in another, within the same country at the same time? Or, why has the rate of wages in the same place been one dollar a day at one period,

never again be enabled to impose such a malignant instrument of fraud upon the community.

Other fortunes which rightly excite jealousy, and which might, perhaps, have been prevented by legal measures, are those which have been made by fraud and by the abuse of trust in corporations on the part of a very few conspicuous or notorious railway promoters and speculators. They need not be named because, fortunately for the welfare of the community, the number of persons who have successfully stolen the property of those who trusted them is very limited; hardly more than one name will come to the mind of any person as the chief exponent of this nefarious class at the present time.

But in regard to such persons it may be said that they are in the nature of monstrosities; they are the spawn of a corrupt period; in one way or another, the man who corrupts a court will be abated in some way as a public nuisance, if death does not fortunately remove him, or ruin does not overtake him.

The great fortunes of those who have fairly earned them by their capacity to direct and use great masses of capital in the most efficient way, cannot be a subject of jealousy, suspicion, or distrust. As well might large steam-engines be a cause of distrust and a clamor be raised for the substitution of a number of little ones.

I have endeavored to show how both the rate of wages and the purchasing power of the wages depend wholly upon the abundance, ready distribution, and quick sale of the joint product of capital and labor.

It is now constantly affirmed by certain enthusiasts and sentimentalists, who are sustained by cranks and demagogues, that, inasmuch as all production rests ultimately upon labor, therefore laborers are entitled to the first consideration and the remuneration of capital ought equitably to be subjected to the prior claims of labor.

This extreme position is the exact reverse of the conception of the relations of labor and capital which prevailed during the first half of the present century, when the science of political economy first became a matter of real study. At that time capital received the first consideration and labor was

and two dollars a day at another, at different times? Third, why is it that one true dollar will buy more in one place than two true dollars will buy in another? Why do absolute wages vary, as they do and have varied, in such proportions as are indicated by the rates in money? And why do the rates of wages vary even when the prices of commodities are the same? In answer to such questions as these we deemed subordinate, or subject, we might say, to capital. One extreme position is as utterly false as the other; both are mischievous; but, if injustice is done in either direction, it is the laborer who suffers most and the capitalist who suffers least. Perhaps the greatest measure of suffering to laborers who are nominally free will be caused when capital and capitalists are subjected to unjust restrictions and injudicious discrimination.

The main purpose of this treatise has been to bring into most conspicuous view the great fact that capital is a *force* which may be applied to the increase of *production* and which *promotes abundance* in the greatest measure; but that it is not a substance to be divided, on the division of which the wages of the laborers depend.

Now, every great force requires the most intelligent and careful direction; the greater the force, the greater the measure of the intelligence and care required. For instance, since the introduction of the steam-engine, or the application of gunpowder to the purposes of mining, no force has been applied with such general benefit to humanity as the railroad whereby the products of the richest sections of the world's surface are distributed over the widest area.

So long as the railway service between the East and the West constituted detached sections, several of which existed between Albany and Buffalo, as well as elsewhere between New York and Chicago—each section being worked under a different administration more or less effective—the general service was ineffective and costly.

It required a man of positive genius in the use of capital and of the greatest administrative power to bring into effect the consolidation of this single line.

It matters not what the motive of the late Cornelius Vanderbilt may have been. It matters not what may have been the motives of those who consolidated that most wonderful organization of all, the Pennsylvania system of railways. It matters not what may have been the motives of those who have laid out the several great systems which are scattered over the country, since Vanderbilt set the example and led the way. The general result of all this work has been a reduction of the railway charge for moving merchandise throughout the United States to the lowest possible point consistent with leaving any

are often answered with the orthodox expression : " Supply and demand determine such points." But this is no conclusive answer until we know under what law the supply has been assured, and under what law the demand exists. These terms, *supply and demand*, are commonly used as if each were absolutely certain to induce the other; but such

incentive of profit sufficient to induce the great masters of the subject to continue their work.

This work is not that of the laborer in the sense in which that word is used by so-called labor reformers. It is not labor in the common acceptation of the term, yet it is an effort of the human mind of such a quality that except capital had thus come under the control of these men all the efforts of laborers would have utterly failed to promote the general welfare. The farmers of the West would have "smothered in their own grease," and would have continued to burn their Indian corn for fuel, while the workman of the East might have starved or would have been compelled to labor long and arduously on the sterile soil of New England, in order to obtain a mere subsistence.

The true function of capital and of the capitalists is of the utmost beneficence. It cannot be exerted in the present condition of the world except by way of the ownership of land and of capital, subject to the limitations and to the duties which are implied by existing laws. That the relations of labor and capital may be measurably changed and perhaps improved by changes in legislation especially in respect to taxation, may not be denied ; but the fundamental principles of individual ownership subject only to the right of eminent domain and to the payment of taxes are essential to that abundant production and ready distribution which makes for the general welfare.

As human nature is now constituted the individual control of capital is essential to its adequate use. Corporations are of the nature of artificial persons, and even they never succeed unless there is some one man capable of becoming the head or chief officer, sustained by as many able assistants as the case requires.

Even the successful co-operative shops in Great Britain exert the closest competition in purchasing their goods and pay very high salaries to the persons who do this part of their work—else they would surely fail. Every co-operative factory is under the personal control of a well-paid superintendent.

"The tools to him who can use them." Capital is a tool which cannot be used except to the mutual benefit of capitalist and laborer. Service for service is its necessary law—the only open question is the ratio which each service bears to the other, and, if my observations are sustained, the law of competition is that the ratio of profits diminishes while the rate of wages steadily increases.

is far from being the truth, except it may be after a long interval of time. Capital may become so effective by the improvement of the machinery in which it consists that a few laborers may be able to supply an article of the utmost necessity in such rapid and excessive measure as to keep the quantity beyond the purchasing capacity of those who need it; the need may exist, but the demand—that is to say, the purchasing capacity—is limited not only by outside conditions, but by personal mental capacity and manual ability of consumers. We may assume, for instance, a community consisting of cotton growers, who raise and pick cotton as a handicraft, and of cotton spinners and weavers who have, also, spun and woven the cotton fibre as a handicraft upon spinning wheels and hand-looms. These two classes now exist side by side in the mountain sections of the South. Up to a given date these two sets of persons may have exchanged services with each other in the ratio of one spinner and one weaver to four growers of cotton; or, in order that we may be able to eliminate those who are displaced by an improvement in machinery, we will assume greater numbers; say in the ratio of one hundred spinners and weavers to four hundred growers. But suddenly capital, in the form of a cotton factory, takes the place of hand spinning and hand weaving; the capacity of a single person operating the machinery of a modern factory being sixty- to one hundred-fold the capacity of a hand worker, and the outside market for the cotton fabric being only among the cotton growers, one hand in the factory exchanges with them, taking their cotton and furnishing them with cloth, and ninety-nine hand spinners and weavers are displaced. They may know no other art. They demand cotton fabrics to cover their nakedness, but they can no longer exchange cloth for cotton. The cotton growers may be able to increase their product in some measure, but they cannot or will not exchange with the hand spinners and weavers when they can exchange on better terms with the factory. The cotton growers and the factory operative may each have more than they had before, and may each prosper; but until the ninety-nine hand spinners and weavers who have been displaced can qualify themselves to do some other service for the cotton growers, or until the cotton growers have developed a want for something else than hand spinning and weaving, there may be no equality in the distribution of the greater supply of cotton fibre and of cotton fabric; there may be want in the midst of plenty. The hard and fast rules of supply and demand must therefore be varied according to the capacity of the persons on whose wants supply and demand are predicated. We heard a great deal about over-production during the long depression between 1873 and 1879, and we are hearing the same cry of over-production at the present time of depression in 1884. Why is this? Over-production simply means an excess of food, fuel, and means of shelter; in other words, it means *supply* of capital. It cannot be said that the people of this country all have so much food, fuel, and shelter that there is no demand for any more. On the contrary, want exists; the need is urgent, but the *demand* does not become potential because something is wanting to bring supply and demand to the terms of an exchange. It takes two to make an exchange. One may have what the other wants, but if the other cannot serve the one, both suffer—one from over-production, the other from under-consumption.

We may perhaps find a clue to this apparent paradox by a consideration of one single branch of industry—to wit, the construction of railways. A railroad is, to all intents and purposes, a product of handicraft. The work done in the construction of a railroad mainly consists in positive, direct human labor, in levelling the way, filling up the valleys, piercing the hills, working in mines and in blast furnaces. Every mile of railroad added to our existing measure stands for the work of about fifty-six men, mostly *common laborers*, working one year. In 1882 we constructed over 11,500 miles of new railroads. In 1884 we shall construct less than 5,000 miles. More than 400,000 *common laborers* have been discharged from work by this change in this one branch of constructive enterprise. They want food, fuel, means of shelter, and clothing now as much as they did in 1882; they represent need or potential demand. Over-production, on the other hand, represents supply; but until other work within the capacity of *common laborers* is found, the wants or demand of these men will not be met, and the over-production or excess of supply will not be consumed. The final end of such a condition is, of course, that pauperism ensues unless an adjustment of labor can be made, and the over-production or excess will then be distributed by the noxious method of alms-giving or State aid. The only true remedy is to develop the individual capacity of each *common* laborer and to render him capable of performing more than one kind of service. To use a Yankee expression, we must evolve "gumption," which is a purely personal quality, in order that there may be neither over-production nor under-consumption.

Let us now return to the direct question : *What makes the rate of wages?* I will now challenge your attention by submitting certain paradoxical propositions which I will presently prove by examples. Although subject to exceptions and to temporary interruptions, they take the form of rules of substantial and uniform application if time be given them to work. In any given country like the United States, where the people are substantially homogeneous, where the means of inter-communication are ample, where there are no hereditary or class distinctions, and where there is no artificial obstruction to prevent commerce, high *rates* of wages in money will be the natural and therefore necessary result of low *cost* of production in labor. That is to say, the two forces of capital and labor being combined in the production of any given commodity, the greatest quantity of that commodity will be produced where the conditions are most favorable and where the least number of persons is therefore required to do the work.

To that point, the best workmen and the most adequate capital will surely tend. This product, whatever it may be, will then fall into the general market of the country, to be converted into terms of money by sale, and will there meet other commodities of like kind which have been produced elsewhere under less favorable conditions or by less skilful persons, with the application of less adequate capital, *i. e.*, poor machinery. That portion which has been produced under the best conditions, will therefore be the representative of the work of the smallest number of persons; and that which is produced under the least favorable conditions, of relatively the larger number of persons. Equal quantities from each source being sold, the sum of money recovered from the sale will be the same, and it will of course yield on the one hand to those most favorably situated, large profits and high wages to the small number employed; and on the other hand small profits and low wages to the larger number less favorably placed. These relative conditions may continue for very many years, as it is not easy to change the place either of capital or of large forces of laborers. *All* will not go to the most favorable place, because there are many other things than mere money which control the disposition of population. For instance, I have given some figures relating to the production of wheat on the great plains of the far northwest. The wheat there produced is greater in quantity in ratio to the capital and to the number of laborers employed, than in any other part of this country, and wages are very high in the harvest season; but it does not follow that every person who has been engaged in raising wheat in Central New York will leave his farm, whether he be owner of the farm capital, or laborer. There are many conditions of life in Central New York which will keep men there in preference to migrating to Dakota, even though both profits and wages be less. Hence it follows, that although the *total* production of any given thing may not be concentrated at the very best point, it will yet be found to be true that where the conditions are the best, the cost measured in terms of days of labor will be lowest, and the wages measured in terms of money per day will be the highest; the high money wages being the necessary consequence of the low labor cost. Conversely, low rates of money wages are the natural and necessary result of a high labor cost of production. This rule mainly affects such products as are made by handwork, or which of necessity remain handicrafts, *i. e.*, work in which the hand is assisted only by very simple tools of which each operation is guided by the hand. In such cases both the materials worked upon and also the product may bear a very high price; but the work upon them, not being aided by effective machinery, the quantity of labor will be very large, and the result of the sale may therefore leave but a very small sum to be divided among very many laborers after the cost of materials has been set aside. All mere handicrafts are quickly overcrowded, except such as call for artistic or original power of design. For instance, after the pattern is drawn it takes merely manual dexterity to make Brussels lace. The material which is used in this branch of industry is fine and costly cotton thread, which is converted into lace by hand without the aid of any machinery whatever, but merely by the use of two or three simple tools; the lace-makers of Brussels are among the poorest of the poorer classes of European operatives. They work at the very lowest rates of wages, which will barely keep them in existence, but their product is of very high cost in money. The very best Lyons silks and German velvets are other examples. They are made upon hand-looms of the most primitive kind. Beet-root sugar is another example. Beets require constant hand work in weeding. We cannot afford the time or labor for such work so long as we can exchange wheat raised by machinery for money and with the money buy our sugar. In all handicrafts the quantity of labor is very great, but even at the high prices which such products bring, the total sum of money recovered from the sale leaves but a very low rate of wages to be divided among those who have performed the work.

It thus becomes very apparent that the rate of wages must be determined by what the product will bring in the market, from which must be deducted materials and profits. The total annual product may be converted into a lump sum of money, which will represent the combined result of the sale of each particular part of the annual product, each part of which has been separately converted into a definite sum of money by sale. From the gross sale of the whole the general rates of wages and profits are, and must be derived; and from the sale of each particular part the rate of wages and the rate of profit on that part *i. e.*, in that branch of industry, must be measured and defined.

So long as we consider the total product of the United States as a unit or single subject of division, the conception of that division may be limited to the two objective points of profits and wages.

Reverting to the algebraic formula, a simple statement serves: x being the value of the annual product, the formula is: $x - a$ (profits) $= b$ (the sum of the wages of all persons employed). But when we take up any special art the proposition becomes a very complex one, and it is extremely difficult to separate the various elements of a given cost, except by the measure in money in which such elements of cost are usually expressed. Each part of the work must be considered separately in order to prove that the rate of wages of each body of workmen who are engaged in each part of the work constitutes a remainder over, and is a result or consequence, rather than an element or measure of cost, as it is usually considered.

We may perhaps solve this problem by an example, and for this purpose a cotton fabric may best be taken, because it is an example of production to which the highest art in the application of machinery is necessary in one department, as well as the lowest-priced manual labor, but little aided by machinery in another.

The elements of a cotton fabric are:

1st. Cotton, including the profit of the cotton farmer, the wages of the cotton laborer, and the wear and tear of the capital or tools used in the production of the fibre.

2d. Other materials, which need not be considered separately, as the same principles which govern the supply of cotton also govern these.

3d. The transportation or movement of the cotton to the factory.

4th. The wear and tear or depreciation of the factory resulting both from use and from the invention of better machinery.

5th. The wages or earnings of those who do the work.

6th. If taxes are levied upon machinery, the capitalist will also assure himself that he can charge the taxes as a part of the money cost of the goods before he builds the mill, and thus distribute them upon consumers, but they do not of necessity enter into this consideration.

With respect to cotton, no attention need be given to any assumed value of land in the southern United States, considered merely as land. The area of cotton cultivation has never yet equalled three acres in one hundred of the area of the cotton States, and if the same measure of intelligence were applied to cultivation in all the States which was given to cotton production by the late Farish Furman, of Georgia, the whole commercial cotton crop of the world, including that of the United States, India, Egypt, and South America, could be produced on one fifteenth part of the area of the single State of Texas.

The price of cotton, therefore, yields profits to the farmer and wages to the laborer; as time goes on, the two are becoming more and more identified. The price of the cotton is determined by competition in the great markets of the world—in Liverpool, Havre, and New York. When the cost of transportation has been set aside and the profit of the cotton farmer has been realized the remainder over, although it is but a small sum per pound, yet suffices to pay the laborers upon the cotton farms of the United States the highest *rate of wages* earned by the cotton cultivators of the world—a far higher rate than can be attained by the ryots of India, the fellahs of Egypt, or the peons of South America. The purchasing power of the wages of the negro of the southern cotton field is also very high when measured by his wants; he prefers bacon and corn—" hog and hominy "—with a little molasses, to any other food; his week's ration consists of three and a half pounds of bacon and one peck of meal, and this can be furnished him at fifty to seventy cents per week, according to the season and to the abundance of the western crops, or at seven to ten cents per day. The food of the rice-fed races of India costs less nominally, but if consideration be given to the force concentrated in and represented by the food, there is probably no other laboring force in the world which can be subsisted at so low a cost, either measured in labor or in money, as the freed negroes of the South.

The price of raw cotton being thus determined, the place at which it may be converted into cotton cloth must next be determined. Into this question many conditions enter:

1st. The use of water or steam power.

2d. Climatic conditions.

3d. The density of the population and the capacity of the separate members of the population to do the work.

4th. The proximity of the factory to the market in which the principal demand exists.

5th. The consuming power of the community in the midst of which the factory is placed, and their ability to buy the products for which the cotton fabrics made in excess of their own wants are exchanged.

Omitting all consideration of fine cotton fabrics, which perhaps depend upon the relative or constant humidity of the atmosphere in the choice of the place where they are to be made, but which are of little relative consequence in the supply of clothing,—and limiting our attention to pure cotton fabrics of heavy or medium weight, which constitute the most important portion of the supply of such fabrics, it appears that the lowest cost of production has been attained in some of the principal factories of New England, of which the specific data are given in the appendix. The fabrics made in these factories meet those of other countries in China, India, Africa, and South America, and are there sold in competition. The price received has thus far sufficed to defray the cost of the materials, the transportation of the cotton from the southern field to the northern factory, the heavy local taxes, a reasonable rate of profit to the owners, and the remainder over has sufficed to give the operatives the highest rate of wages earned in this art in any part of the world. Whether this superiority can be maintained by New England in competition with the Piedmont section of the Southern States is now considered an open question by some observers. In this treatise it will suffice to call attention to two facts by which the propositions herein submitted are fully sustained.

1st. That in this art the rate of profit in a given product has steadily diminished, and the rate of wages (or of the remainder over) has as steadily increased.

2d. That in the most important division of this art, to wit: the manufacture of coarse and medium fabrics from cotton unadulterated with clay, the highest rate of wages (or remainder over) is realized where the cost of production is lowest, *i. e.*, in New England.

In treating this subject it matters not whether this result has been reached by means of a protective tariff, or in spite of one. It is admitted that special rates of wages in a particular art may be raised by the exclusion of a foreign product of like kind, so long as the price of the domestic product is maintained above what it would otherwise be; but this is exceptional. I have selected examples of products of which the price is determined both by domestic and by foreign competition, in order that the main question may not be confused by any prejudice for or against any special policy. Reference will be made hereafter to the conditions under which the policy of protection may or may not be expedient.[1]

---

[1] In this connection the writer may venture to express an opinion as to the place in, or section of, the United States where the cotton manufacture will be gradually concentrated.

It has been submitted that the most ample capital and the most skilful labor will tend to the most favorable place, because at that place the remainder over of which wages consist will be the greatest proportion recoverable from the sale of the product.

Steam having substantially displaced water as the motive power of the factory, the climatic or atmospheric conditions in which the cotton fibre can be most successfully spun and woven have become perhaps the most important elements in determining the place of conversion. In England there is a steady and constant trend of the spinning mills to the points where the deposition of moisture is most uniform, and where the humidity of the atmosphere is most constant. There is scarcely a spindle left in Manchester and there are eleven million spindles in Oldham, a town which has grown from insignificance to this importance in a very few years. It is about 800 feet above the level of the sea, on the edge of the level moors at a point where the deposition of moisture is constant. In this country it may perhaps happen that cotton spinning will be concentrated more and more along the coast of the southeastern part of Massachusetts, in Rhode Island, and along the coast of Connecticut, where the influence of the Gulf Stream is most apparent, and where cotton and fuel can be laid down at the least proportionate cost of transportation. It will be observed that in the annual expenses of families living upon an income of $500 to $800 per year the cost of mere subsistence is sixty per cent. of the whole expenditure. In the section designated the staple articles of western food—grain and meat—can be delivered at a cost of $5 per ton for over 1,000 miles of distance, and one ton suffices for a years' ration of grain and meat for four or five persons. On the other hand, this section has a positive advantage over almost any other in respect to groceries and in the supply and preservation of vegetables, while its distance from the cotton field is fully offset by its greater proximity to the principal markets for goods. The colder climate of winter gives a necessary stimulus to industry, and is more readily qualified than the excess of heat in the southern summer. Hence it may happen that at this point, or in this section, the highest wages will always be the remainder over from the manufacture and sale of staple cotton fabrics.

In this section the population will also be likely to remain more dense, and also more capable of great diversity of employment and subdivision of labor. These are very important considerations, since the margin of profit is becoming

Wages are held to be a consequence—a result—a remainder over after capital has received such profit as will have induced it to undertake the work; *the rate of wages cannot therefore be considered a true measure of the cost of production.* Wages are a consequent result, and their measure or rate is, and must be, determined, in the long run, by what the product will bring, and not by what the capitalist may either promise or be willing to pay for a given time. He may not be able to forecast the future in such a man-

---

less and less. It may almost be said that in all the great arts the profit is found in the utilization of the waste or of the secondary product of the factory, and in the facility with which the machinery can be kept up without the necessity of maintaining a large force of spare hands under constant pay. Hence the isolated cotton mill, which is far away from the paper mill on the one side and the machine shop on the other, is at a relative disadvantage which tells against it in the close competition under which a quarter of a cent on the yard of cloth is equal to four or six per cent. on the capital invested. This tendency of particular arts to become fixed in particular places calls for more attention than has yet been given to it, in order that the reasons may be fully comprehended and their influence on wages considered.

It would be a matter of curious interest to study the forces or influences which made gloves the chief product of Gloversville in New York, and gave the town its name; why card clothing is made chiefly in Leicester and Worcester, Mass.; why men's heavy boots are made in Spencer and Brookfield, and women's boots and shoes in Lynn; why brass work of certain kinds is conducted so largely and exclusively in a few towns in Connecticut; etc., etc. There are, of course, very obvious reasons why primary work of many kinds should be found in special places, but the reasons for the concentration of secondary work are not so plain, and a study of the causes might yield most valuable results, especially in their effect upon the remainder over which makes the rate of wages in these arts.

The time has been when fine cotton yarn has been spun in England, sent to France to be woven, to Germany to be dyed, and brought back to England to be sold. The best flour of Minneapolis is even now in some small measure sent to London to be baked into biscuit, and is brought back to Boston and New York to find a market. If profits and wages were not recovered from these movements in greater measure, they would not occur. What are the subtle causes of such commerce?

ner as to be able to carry out a single promise which he has made in advance of the sale of his product. The *sum* but not the rate of the wages in any given quantity of products may serve as a means of comparison of the money cost when persons who are engaged in the same branch of business desire to compare their conditions; but the rates of wages constitute no measure of comparison unless the conditions under which the work is done,—that is to say, unless the quality and kind of machinery, the materials used, the advantage of position, the hours of labor, and other elements of the real cost, are absolutely identical.

I have said that in a country which is inhabited by a homogeneous people, the rate of wages will be highest where the conditions of production are most favorable, because the quantity or intensity of the labor will there be least and the product will there be greatest. In like manner when exchanges are made between two different countries, each country will exchange with the other some portion of its own product, which it can make under the most favorable conditions, or in excess of its own needs. The two products being each converted into terms of money will be exchanged as equivalents, without any regard to the proportion or quantity of labor which each represents. We may exchange one day's labor in a Lowell factory in the manufacture of drills, for one hundred days of labor in China in the preparation of tea. It matters not what the rate of wages of the Lowell operative had been, or what the earnings of the Chinamen handling tea had been; their product is converted into terms of money, and is exchanged at certain prices which represent a given number of yards of drills for a given number of pounds of tea. Each is an equivalent to the other. No one asks what the rate of wages or the quantity of labor in each has been. The wages are the result, not the antecedent.

When the exchange is continued—it proves that each party makes a profit by the transaction. The Lowell operative could not have produced the tea, the Chinaman could not have produced the American drill; when the exchange is made, the tea sells in America for more than the equivalent of the drill there, and the drill sells in China for more than the market price of the tea there; therefore there is a certain sum of money, or result of labor expressed in terms of money, to be divided among the laborers in each country, in excess of what there would have been had not the exchange been made. The final result of the labor of the Lowell operative is the number of dollars which the tea brings, less the cost of transportation; that sum is more than the drills would have brought at home, else they would not have gone to China.

Try this on a little larger scale. We now import into the United States, annually, materials which are free of duty to the value of $200,000,000, and we exchange for them, at this measure in terms of money, the surplus of our cotton which we could not now spin ourselves,—the surplus of our oil which we could not now burn ourselves,—and the surplus of our wheat which we could not now eat, even if every man had every day all the bread he could possibly consume. What we send out is our surplus, our excess, a part of our *over-production* which could not be converted into terms of money at any price, or which would have reduced the price of the whole product if it were retained; if retained at home it would yield nothing to divide in terms of money as the equivalent of such excess, among those who did the work. But the substances for which we have exchanged this excess having been brought into the country where they do possess a value of $200,000,000 or more, there is that additional sum to be converted into terms of money and subdivided in profits and wages. In the use of this foreign material, much of which enters directly into the work of domestic manufactures, all wages are therefore, by so much higher than they would have been otherwise. There is so much more to be divided in terms of money, because so much has been added to the quantity of things which could be used; while the cotton, oil, and wheat sent out from the country could not have been used. Now, it matters not what may have been the rate of wages paid in the production of the cotton, wheat, or oil; and it matters not what may have been the rate of wages paid in raising the wool of Australia, in making the tea of China, or in saving the hides of South America. We may receive the work of ten men for one day at twenty cents a day, for the work of a single man working one day for two dollars. By so much as the quantity of labor in our exportable commodities is less than the labor in those which we import, will the rate of wages be higher to our home labor as the necessary result of the exchange, because so much additional substance has been added by import from abroad to the quantity of things for which a home market could be found. This import has been received in exchange for home productions, for which there is no market, because they are in excess of home wants. There can be no continuous commerce unless there is a continuous service or profit to both parties.

It follows that the nation which has diminished the quantity of human labor in greatest measure by the application of machinery, produces goods at the lowest cost, and by exchange with the hand-working nations, who still constitute the majority of the people of the world, are, by way of such exchange, enabled to pay the highest rate of wages in money, because their goods are made at the lowest labor cost. This is the secret of English commerce.

The rates of wages are higher in England than in any country with which she makes large exchanges, except the United States. She buys largely from us in spite of our higher wages, because by way of high wages we make grain, cotton, meat, oil, and many other articles necessary to her use at a lower cost in money than any other nation.

Having thus attempted to present the principle at issue in this matter, let us now consider its application. The only problems of any great importance which are now presented to the people of this country for their determination, consist of the various problems in regard to the collection of the revenue, to the banking system, to the quality and kind of coin which shall be a legal-tender in the settlement of debts, and other fiscal questions. The tariff, the currency, the banking system, and the coinage are the only political questions of any moment. Fortunate for us that it is so, and that we are free from the complications of other countries. Strange it is, and true it is, that the most difficult political question to be dealt with by the people of the United States is, *how to get rid of a surplus revenue.*

Neither one of these problems can even be stated without immediate reference being made to their bearing upon the rates of wages of the people of this country.

Aside also from questions of revenue, banking, and coinage, the relations of men to each other cause discussion,—the hours of labor, the respective duties and rights of employers and employed, competition and coöperation, and all the other subjects which are customarily summarized under the general term of "the labor question." Not one or all of these questions can ever be discussed without an immediate consideration of the rate of wages. In every speech, in every essay, and in every conversation by the way, upon any of these subjects, the *rate* of wages comes at once to the front, and, as a rule, one or the other of the following propositions is almost invariably assumed, all of which are the very reverse of being true, and all of which are inconsistent with the law of wages which I have attempted to propound. All such discussion serves but to confuse the mind, simply because no distinction is made between the rate of wages and the sum of wages, and because it is assumed that all laborers or operatives are equally efficient.

I again desire to express the hope that the form of these propositions may not prejudice any one, be he an advocate of protection or of free trade. The so-called principle of *laisser faire* is by no means implied in this treatise. The welfare of laborer and capitalist rests upon many other conditions than the rate of profits or wages, but the forces which determine these rates must be fully considered before any intelligent discussion of any social question can be undertaken. It is to these forces that I have endeavored to limit this treatise. I will state these fallacious propositions in order, as follows:

POPULAR FALLACY NO. 1.

The cost of production of any given article can be ascertained by finding out and comparing the *rates* of wages paid in its production in different places, here or elsewhere.

POPULAR FALLACY NO. 2.

Low *rates* of wages are necessary to low cost of production; high *rates* of wages can only be paid consistently with high cost of production.

POPULAR FALLACY NO. 3.

Inasmuch as laborers work for wages, wages enter

directly into the cost of production, therefore cheap labor can only be assured by the payment of low *rates* of wages.

POPULAR FALLACY NO. 4.

An employer must of necessity be able to hire laborers at low *rates* of wages in order to make goods at low cost.

Now if one asks any employer which workman is the first one to be discharged in a period of depression,—the workman who, being employed by the piece, earns the lowest rate of wages for himself, or the one who earns the highest,—unless some other question than the mere cost of goods enters into his consideration he will reply: " Why, the poor workman will be discharged first, of course,—he who earns the lowest rate of wages." Each employer understands perfectly well in his own business that the cheapest man,—that is, the man who does the most work for the least money is the one who works the greatest amount of machinery with least stops, *i.e.*, the most effective workman ; in manual labor it is the strongest; in a handicraft it is the one who possesses the greatest manual dexterity; in the operation of machinery it is the one who understands the machine best and can get the most work out of it. The very man who may have taken part in a discussion in which he has assumed that the popular fallacies which I have recited are unanswerable truisms, will never conduct his own business consistently with them, and if he did he would be sure to fail sooner or later.

The true cost of any given article is the quantity of labor or the human effort expended in its production ; now, if we consider a human being as an automatic machine, similar to any other mechanical power or force, the true cost is the quantity of food and fuel expended in the conversion of a given amount of material substance into human force.

How true this is has been proved by Brassey in his comparison of the cost, even in money, of the labor of the English navvy as compared to the Hindoo or any other of the rice-fed people of the world. This human effort is measured or converted into terms of money, and it is the sum of the wages, not the rate, which constitutes the money cost; to this sum the rate of wages may bear a large or a small proportion. Wages in money are the instrumentalities for procuring food, fuel, and shelter; and the worker is practically the more effective, the more money he can earn, or, in other words, the more money he can spend in a judicious manner for a good subsistence. The English navvy may be instanced again as being worth twice as much, either in the measure of his work, or by converting the measure of his work into wages, as the rice-fed coolie. He earns more, he spends more, he eats more, and he does more than double the work. Therefore, although he attains a high rate of wages, the result of his labor will be a lower cost of production. Again, the skilful weaver who can tend six looms, and keep each loom moving, being paid by the piece or according to the quantity of cloth woven, earns higher wages than the unskilful weaver who only tends four looms, and has one stopped a large part of the time; the sum of the wages of the six-loom weaver is the least in proportion to the quantity of cloth produced. The high wages represent the low cost.

Not very long since, a German steamer, on the way to New York, was very much damaged, so that very extensive repairs became necessary. It was decided to do the work of repairing in New York, as it appeared difficult to send her back to Bremen; but the agents were instructed to report in Bremen, day by day, the number of men employed and the rates of wages; which report they made. When the first report was received in Bremen, a telegraphic message was returned, ordering the steamer back to Bremen for the completion of the repairs, for the reason that the owners of the line said that they could not afford to pay such high rates of wages, being well assured that the cost of repairs would be more than what they would of necessity expend in Bremen. But it was too late; the work had been begun and it was necessary to finish it in New York. When the final account of the *sum* of wages was sent to Bremen, it proved to be a less amount than the same repairs would have cost in Bremen. Since then there has been no reluctance to repair these German steamers in New York.

Again, the rates of wages may be precisely the same in two factories in the same place, and yet the cost of production will vary so much that one mill will prosper while the other will fail, because the quantity of product will vary, and the profit or loss of any textile factory rests mainly upon the quantity of yarn spun and of the goods woven. There may be many reasons for this difference: in one mill the machinery may be old, in the other new; in one the material may be well selected, in the other badly; in one the goods may be well sold, in the other badly sold; in one the goods may meet the fashion, in the other they may be out of date, although better in quality. Under all these varying conditions, the source of wages being the money produced by the sales, high wages may have been paid consistently with low cost of production in one factory; and low wages may have been paid, notwithstanding the high cost of production, in the other; or, if the cost of production be the same, the goods of one mill being well sold and those of the other ill sold, the sum left to be divided might amply suffice for high profits and wages in the one case, and be deficient in the other. Thus, difference in management will alter results, in the same place, at the same time, in the use of similar machinery. The same management will yield different results, both in profits and wages, on different machinery. The same management and similar machinery will yield high wages in one place, and the reverse in another, at the same time, because the conditions vary in other respects.

I have submitted these several propositions under the name of popular fallacies. It will be apparent that a very large part of the discussions in respect to hours of labor, in respect to taxation, and to all other matters connected with the so-called labor question, are commonly based upon them, and the common conclusions are as fallacious as the propositions.

A true theory of the source of wages and their actual relation to productive industry is therefore necessary to any intelligent discussion of any of the questions now before the country.

The wage question must be treated from four points of view.

First.—What individual effort is required to earn a given sum of money in a given time?

Second.—What is the purchasing power of that money?

Third.—What are the relative efforts, as well as relative sums of money earned in the form of wages, by those who compete in a given product in the same or in different countries?

Fourth.—What is to be considered in addition to the cost of materials and the rate of wages, in placing the goods produced at the point of consumption?

The fallacies which have been previously submitted may be met by counter propositions, all of which can be substantially sustained; exceptions being readily designated, and the reason for such exceptions being readily found.

First.—The rate of wages constitutes no standard even of the money cost of production; which cost must be made up by adding together the sum of all wages and dividing by the product, in order to establish a unit of cost in money by way of a unit of measure—whether by the yard, barrel, or pound.

Second.—Low rates of wages are not essential to a low cost of production, but on the contrary usually indicate a high cost of production,—that is to say, a large measure of human labor and a large sum of wages at low rates. Conversely, high rates of wages may, and commonly do, indicate a low cost of production,—that is to say, a small proportion of human labor and a small proportionate sum of wages at high rates in a given quantity of product.

Third.—Cheap labor, in a true sense, and low rates of wages are *not* synonymous terms, but are usually quite the reverse.

Fourth.—An employer is not under the necessity of securing labor at low rates of wages in order to make cheap goods, but he may and commonly does pay high rates of wages, for the very purpose of assuring the production of goods at the lowest cost,—that is, in order to be able to sell them on the lowest terms, or "cheap" in the popular sense.

The abuse of the word *cheap* leads to more mischievous fallacies than any other abuse of language. The cheapest labor is the best-paid labor; it is the best-paid labor applied to machinery that assures the largest product in ratio to the capital invested.

If these propositions can be sustained, it may be submitted that the more the capitalist increases his wealth and applies it to reproduction, the more the welfare of the laborer is assured. The competition of capital with capital tends constantly to a decrease in the ratio of the profit of capital to the total production, and of necessity tends also to a constant increase in the rate of wages of the laborer; thereby more than counteracting the tendency of the competition of laborer with laborer to diminish wages.

I will now attempt to prove these apparently paradoxical propositions by one of many examples by means of which this theory can be sustained. It will be taken from the records of the cotton manufacture, not only because this branch of industry is most familiar to myself, but because it was almost the first of those which were brought under the factory system by division of labor, and under this system factory accounts have been kept in the same way from the very beginning.

In 1830, when the first statistics in my possession are dated, the average earnings of all the operatives in a large cotton-mill, who then worked thirteen hours or more a day, and among whom were comprised a much larger proportion of men than at the present time, while the women were older and there were fewer children, were $2.50 to $2.62 per week. The quantity of machinery which each hand could tend was much less; the production of each spindle and loom was less; the cost in money of the mills per spindle or loom much greater, while the price of cloth was at times more than double the price at which it can now be sold with a reasonable profit. The average earnings of all the female operatives in what purports to be the same factory, at the present time, on the same fabric, working ten or eleven hours a day, under vastly better sanitary conditions, both in the factory and in their dwelling-houses, are $5 per week, and in some cases even $6—or more to the most skilful. That is to say, women only now earn about twice as much in ten hours as men and women combined averaged in thirteen hours a little over forty years ago.¹ Between these two dates, subject to various fluctuations from temporary causes, the course of events in this branch of industry has been as follows: A continuous reduction in the hours of labor, coupled with an increase in the earnings per hour; a diminution in the money value of the machinery,— that is, in the ratio of capital to production, coupled with an increase in its productive efficiency; a constant increase in the supply of cotton fabrics per capita, coupled with a decrease in the price; a continuous increase in the purchasing power of gold dollars in respect to almost all articles of necessary subsistence, a few articles only having advanced in price, mainly meat and timber.

In all these points the cotton manufacture is not exceptional, but the same facts can be proved in respect to all other branches of industry where the accounts have been kept upon a uniform system.²

After making all necessary corrections in the data respecting cotton fabrics, on account of the variations in the price of raw cotton, it therefore appears that the apparently paradoxical propositions which I have submitted—the reverse of those which are commonly accepted—are fully sustained.

First.—The rate of wages paid has not been a true measure of the cost of production.

Second.—The lowest rates of wages have been paid when the cost in money was the highest, and the highest rates of wages are now paid when the cost in money is lowest.

Third.—Low wages and cheap labor have not been synonymous terms. That labor has, in fact, proved to be cheapest by which the largest product for each dollar ex-

---
¹ See appendix. Graphical statement of two factories.
² Appendix.—Wages of various kinds compared.

pended was assured, and that has been the highest paid labor.

Fourth.—The employer has not been under the necessity of paying low wages in order to make low-priced goods. The goods now made at the rate of $5 to $6 per week being sold at less than one half the price, in many instances, of those which were formerly made at the rate of $2.50 to $2.62 per week. Not only is the capital in the cotton-mill now less than one half what it was in 1830 even when measured in terms of money, in ratio to the value of the product, but the average rate of profit which capital now rests satisfied with is less than half on each dollar invested what it was in 1830. Hence the competition of capital with capital has increased the quantity of cotton cloth at a decreased rate of profit. On the other hand, the competition of labor with labor has not prevented the continuous rise in the rate of wages, and these wages have more than doubled in the purchasing power of each dollar, by comparison with the cotton cloth in the making of which they have been earned. In respect to some kinds of cotton cloth, such as printed calicoes, the actual weekly wage of to-day will buy four or five times as much as the weekly wage of forty years ago. In this branch of industry, at least, all interests have thus been harmonious. The increase of wealth in the cotton manufacture has been accompanied by a yet greater increase in the welfare of the cotton operative, while both have been accompanied by a vastly greater supply of cotton fabrics, and by their increased consumption at lower and lower prices.

These data have been compiled from the accounts of certain factories which have never become bankrupt—whose stock has never been reduced in its par value, and which have paid a fair average dividend to their stockholders, from time to time, since they were established to the present day. I have taken as examples coarse fabrics, the common wear of the million. During this period, from 1830 to 1884, this branch of industry, like all others, has been subjected to over thirty changes in the tariff; to the suspension of specie payments in 1837 and 1857, brought about by purely commercial crises; to the suspension of specie payments at the beginning of the war, brought about by the imposition of the Legal-Tender Act; to a variation in the price of cotton from five cents a pound to $1.83 per pound; to the weary depression from 1873 to 1879; to several minor commercial crises. They have also been subjected to numerous acts of interference on the part of the State Legislature in the conduct of their affairs. If constant vacillation and change in acts of legislation, in respect to the tariff, currency, banking, bankruptcy, taxation, hours of labor, and other acts which are now deemed of present permanent interest to legislators, could have killed these establishments, they would have long since been very dead. May not this prove that we depend much less upon governments and upon statutes than we think we do? We are almost forced to accept the dogma of Buckle, that the greatest service of modern legislators is to repeal the obstructive statutes of their predecessors.

The same progress and improvement in the condition of the operative have occurred in England during the same period; only the change has been greater there than it has been here, because the English operatives started from a much lower plane and have now nearly attained an equality with the condition of our own in many departments.

We may now recur to the question, What makes the rate of wages? In other words, Why are the average wages expressed in terms of money in the same factory nine to ten cents an hour to-day, against three and a half to four cents an hour forty or fifty years ago, while the rate of interest or profit on capital, when invested in the safest possible securities, is now only three to four per cent. against six, eight, or even ten per cent. then?

In order to bring out the point of this argument with yet greater clearness, having already compared one period of time with another in the same factory, we may now compare one mode of work in this art with another in the same country in two different places, to wit: Let us compare the homespun fabric of Western North Carolina with the factory cottons of New England. It is computed by men who have had much experience, and whose observations are entitled to credence, that there are two or three million persons living in the heart of the United States, in the mountain section of the South, who are still clad in homespun fabrics of cotton and of wool. I have myself been among them, and have examined the conditions of the art of making cotton goods as it there exists. Two carders working with hand cards, two spinsters operating spinning-wheels, one weaver working a hand-loom—five adult persons in all—convert four to five pounds of cotton into eight yards of cloth in ten hours; the cloth is heavy, rough, and unsightly, very durable, and worth in the neighborhood, when sold, about twenty cents a yard. If the value of the cotton be deducted, the five persons might possibly earn twenty cents a day, the total value of this product being $1.60. The capital invested in the hand machine can hardly be computed, because the only thing purchased would have been the two hand-cards; but if the hand labor expended in the construction of the spinning-wheels and hand-looms were computed in money, the whole investment might come to $100. The proportion of capital used, in its ratio to the annual product, would therefore be very small, and the ratio of labor, even at twenty cents a day, be very large. In New England, $5,000 worth of capital, operated by five persons, male and female, averaging each one dollar per day in wages, will suffice for the conversion of three to five hundred pounds of cotton into eight hundred yards of the same kind of coarse cotton cloth; the cloth softer, more sightly, and not quite as durable; when sold as low as even seven or eight cents a yard, yielding money enough to pay for the cotton and other materials, profit enough to pay ten per cent. on the capital, and yet leaving as the result for the wages of the operatives one dollar a day as their share of the product. Between these two extremes every phase of the progress of a century in the art of cotton-spinning and weaving can even now be observed, in a journey of a week, from Boston to North Carolina and back. The small mill, like that of 1828, fitted with old, heavy, slow-moving machinery, still exists, in which twice or thrice as many Southern operatives, working thirteen hours a day, at two thirds the rate of earnings made in Lowell, get off a less product of cloth at a far higher cost. As we journey back toward the North, the mill becomes larger and more effective, until we arrive at the great factories in New England, where the highest wages are paid and the lowest cost of production is assured. The same or even greater extremes may be found by comparing India and China with England; while the cotton-mills of England, when compared with the factories of Germany and Italy, although the machinery may have been made by the same makers, yet show the same rule—a larger number of persons, less effective work, lower rates of wages, and higher cost, as we go away from England to Germany, to Austria and to Italy.

It would therefore appear that wages are a remainder

over from the sale of the product, and are determined by the sum of money which that product will bring in the markets of the world. From this sum of money must be assigned:

First.—A portion or sum sufficient to restore the depreciation of the capital used,—in other words, to keep the machinery in effective condition.

Second.—A sum equal to the average rate of profit on capital invested in the very safest securities, and, in addition to that rate, as much more as is necessary to compensate the owner for the greater risk of one branch of work as compared with another.

Third.—The cost of the materials.

Fourth.—The sum needed to secure the very best administration.

Fifth.—The proportion of the national, State, and municipal taxes which are collected from the consumers of the goods through the instrumentality of the person, firm, or corporation owning the property; which taxes enter into the money-cost of the product and must be recovered from the sales.

Lastly.—The remainder over constitutes the wages or earnings of the laborer, whatever that remainder may be.

Profits, taxes, and wages are therefore alike derived from the sale of the joint product of capital and labor.

Unless one branch of industry yields the average of all branches, due regard being given to the greater or less risk of each as compared with the other, it will not be undertaken; or, if undertaken, it will not long continue to be pursued. Wages therefore are apparently deferred to profits; but, on the other hand, wages constitute *all that there is left*, and under the inexorable law of competition of capital with capital, the profits of capital are constantly tending to a minimum, while the rate and purchasing power of wages are both constantly tending to a maximum. Capital is always ready to take the risk and to become the guaranty or insurance fund for the recovery from sales of goods of higher and higher wages for any kind of skilled labor which is capable of increasing the product of any given quantity of machinery. From the sale of this increased product, in the first instance, capital gains. More of the same machinery is then added, and, as it becomes greater in quantity and more effective in use, the rate of profits diminishes, although the aggregate may increase; in other words, capital secures a less and less proportion of the constantly increasing result, while labor receives all that there is left over. That is, the remainder over is constantly becoming a larger and larger proportion of an increasing product. There are of course temporary fluctuations; but both observation and experience, combined with statistics, confirm this rule both in this country and in England. In other words, the rule laid down by Bastiat is sustained by experience; the aggregate profit of capital is augmented but the relative profit is diminished while the wage of labor is increased both absolutely and relatively.

I had been engaged in this examination and compilation before I even knew that Mr. Robert Giffen was engaged in the same work. His results and my own, covering a period of fifty years, are identical.

Having thus attempted to answer the general question, What makes the general rate of wages? now let us give a few moments to the particular question, What makes the rate of wages higher in this than in any other country? In order to give an intelligent reply to this question, we must treat the annual product of the United States as a whole, and consider only the general rate of wages in this country. In some particular branches of manufacture, or in some hereditary or national arts, other nations may still apply machinery more effectively than we do; and in some special branches of agriculture, such as wine, olives, sugar, and the like, other countries may either possess better conditions or for the time being greater skill. On the whole, however, the people of the United States are in the possession of more ample and varied natural resources, and of the most effective capital in the form of machinery; they are also endowed with greater facility in the adaptation of machinery both to agriculture and to manufacturing; they possess more effective mechanical instrumentalities of distribution by rail and river; they enjoy a continental system of unrestricted commerce between the States; they have a fairly complete system of common education; but lastly, they are subjected to the least diversion of any part of their annual product to purposes *of destructive taxation*,—that is, to the support either of standing armies or of privileged classes. I do not recite our advantages in a boastful way but in order merely to bring out the salient point, that *while other Nations prepare for War we prepare for Work.*

Our only great war has been fought in the interest of labor—in order that labor might be free. It gave such an incentive to invention in the North that all our principal crops increased during this period even though a million men were taken away from their work. It opened the way for the Southern States to such conditions that the South itself is to-day richer and more prosperous than in the palmiest days of slavery.

Our national debt in 1866 was $83 per head of population. It is now but $25 per head, and will soon be wholly paid.

When two simple principles shall have become a part of the common knowledge of the people of the United States, the end of all standing armies in the civilized nations of the world will have come.

These two principles are:

First.—All nations are interdependent, and in all commerce both parties gain in welfare.

Second.—In all arts which are not mere handicrafts high wages in money are the necessary result of low cost of labor of production.

In the grand competition for the commerce of the world which now turns on a cent a bushel, a quarter of a cent a yard, or a fraction of a penny on a pound of iron or steel, no nation which bears the burden of standing armies like those of Germany, France, Italy, Austria, and Russia, can hope to enter into successful competition with England or the United States, when the whole English-speaking people take advantage of their position and serve the nations of the world with goods at low cost, in which all who have joined in the work have made higher wages than can be earned in any of the countries named. The commerce of the army-burthened nations with others will be destroyed by its own restrictions. Nations can only be ruined by their own burdens;—then what may come? Their own resources will not suffice to sustain their armies, but with the burden of their armies upon them they cannot engage in competition with England or America; their product will be small and insufficient; their wages very low in their rate, barely capable of buying enough to sustain life—if even for that,—while their cost of production as a whole must be very high.

It is difficult to foresee the course of events. These armies are as impossible to be disarmed as they are incapable of being sustained, without revolution and destructive war. What will be the end no man can tell!

In contrast with these adverse and costly conditions, the English-speaking people may well rejoice in the relative freedom of Great Britain and the absolute freedom of the United States.

In addressing the British Association it may not be unsuitable to call attention to the position of the United States, provided it is not done in a boastful spirit. In dealing with the potentialities of the future it is almost impossible to prevent the imagination from running riot, but since the Chairman of our Section, Sir Richard Temple, has spread before you in his address the magnificent picture of the British Empire, I may perhaps be permitted to dwell upon the resources of the United States, and by analogy, of Canada also, in a few paragraphs. With respect to my own country, I may venture to say that in addition to the advantages I have recited our taxes are, on the whole constructively expended. The necessary result ensuing from our conditions is a larger annual product in ratio to the number of persons employed in making it, measured either by quantity, or, when brought into competition with the world, by price or the sum of money which is received for it, than can be elsewhere attained. It is also, as a rule, of better quality, because of the more intelligent methods applied to its production. If we consider production as a whole, our annual product comes into competition for sale, with other products of the world of like kind, and its price as a whole, is determined, directly or indirectly, by this world-wide competition. From this determination of its price, its value is converted into terms of money. Quantity and quality alike tend to increase the sum of money recovered from the sale, and this sum of money is the sum which is to be divided between capital and labor. Large general profits and high general rates of wages are the necessary result.

It is therefore proved to have been absolutely true in

this country that, in proportion to the increase of capital, the absolute share of the value of the annual product falling to capital has been augmented, but its relative share has been diminished; while, on the other hand, the share that has fallen to labor has been increased, both absolutely and relatively. The generally high rate of wages, expressed in terms of money, in the United States, is the necessary consequence or result of the generally low labor cost of production,—that is, of the smaller quantity of labor by which the production is assured ; which less quantity of labor suffices because of the application of the most effective machinery, *i. e.*, of capital, to the work.

Let me give two or three salient examples proving this rule. Man does not live by bread alone, but bread is the staff of life. What people gain their bread with so little exertion of human labor as the people of this country? If we convert the work done in the direction of machinery upon the great bonanza farms of far Dakota into the yearly work of a given number of men, we find that the equivalent in a fair season, on the best farms, of one man's work for three hundred working days in one year is 5,500 bushels of wheat. Setting aside an ample quantity for seed, this wheat can be moved to Minneapolis, where it is converted into 1,000 barrels of flour, and the flour is moved to the city of New York. By similar processes of conversion of the work of milling and barrelling into the labor of one man for a year, we find that the work of milling and putting into barrels 1,000 barrels of flour is the equivalent of a man's work for one year. By a computation based upon the trains moving on the New York Central Railroad, and the number of men engaged in the work, we find that 120 tons, the mean between 4,500 bushels of wheat and 1,000 barrels of flour, can be moved 1,700 to 2,000 miles under the direction of one man working eighteen months, equal to one and a half men working one year. When this wheat reaches New York City, and comes into possession of a great baker, who has established the manufacture of bread on a large scale, and who sells the best of bread to the working people of New York at the lowest possible price, we find that 1,000 barrels of flour can be converted into bread and sold over the counter by the work of three persons for one year. Let us add to the six and a half men already named the work of another man six months, or half a man one year, to keep the machinery in repair, and our modern miracle is that seven men suffice to give 1,000 persons all the bread they customarily consume in a year. If to these we add three for the work of providing fuel and other materials to the railroad and to the baker, our final result is that ten men working one year serve bread to one thousand.[1]

---

[1] It may not be assumed from this analysis of the production of wheat upon what are known as the great "Bonanza Farms" of the Northwest, that any inference is to be drawn from these facts either for or against the large holdings of land as distinct from small farms or "peasant proprietorship" so called.

If consideration be given to the kind of crop which is to be raised, it will be apparent that a certain proportion of the products of agriculture may rightly be raised upon the largest allotments of land to which machinery may be applied in the greatest measure, by which method the largest production will be assured at the least cost.

Wheat is essentially a crop of this kind. It contains the maximum of nutriment in the least bulk. It can be moved over long distances at low cost, and it is a prime necessity of life. It may, therefore, well be produced in largest quantity at the lowest measure of cost, even though this method may for a time injure the condition and impair the prosperity of the small farmer who cannot adopt machinery in so great a measure, or who has not the capital necessary for extensive cultivation.

Maize or Indian corn, on the other hand, containing less value in the same bulk, may well be raised upon smaller allotments of land nearer the places of consumption, if it is to be used in the form of meal; but maize may also be considered one of the crops subject to the application of large capital, and to being raised in the most economic manner on large farms when it is to be fed to cattle or hogs, and thus concentrated into a removable form.

Again, iron lies at the foundation of all the arts. At an average of 200 pounds per head in the United States, the largest consumption of iron of any nation, we yet find that the equivalent of one man's work for one year, divided between the coal mine, the iron mine, and the iron furnace,

But although the wheat and corn crop constitutes so large a factor in the subsistence of the people, there are yet very many other products of agriculture which can only be raised in part by hand labor or with less application of machinery, and upon small farms more economically than they can be upon large ones. Hence it follows that in districts like the central part of the State of New York, which was formerly the great centre of wheat production of the United States, as soon as the competition in the sale of grain of the great Western farms began to be severe, the land being under no restriction either of lease or settlement or other artificial condition, was immediately converted to other crops, such as fruit and vegetables, which will bear transportation over short distances only, or of seeds and the like ; while the land in closer neighborhood to the great cities, which under former conditions and in the absence of cheap transportation was of necessity devoted to the coarser or more staple crops, is now devoted to market gardening.

Thus it has happened that while the large farmers prosper the small farmers prosper yet more, not being under the necessity of applying themselves to a few coarser staples, but adapting their land to any demand which may happen to exist in their immediate neighborhoods.

The production of wheat in the central part of New York is about as large as it ever was when it was the great wheat centre of the country, yet it is now a very insignificant factor in wheat production, and the farmers in this section have attained vastly greater prosperity by diversity in their production, and by the application of improved tools combined with hand labor, than they ever obtained under the former method.

The secret of success in agriculture, as in many other matters, therefore, lies in the freedom of the land from the artificial restrictions of leases, settlements, and the like—by which English land is now so much encumbered, and the reason why the agriculture of the Eastern and Middle States has advanced in method and prosperity in the face of Western competition, is to be found in the absolute freedom in the purchase, sale, and use of land, which is the rule in this country. Land is itself a tool or instrumentality, and under our laws and customs the tools ultimately fall to him who can use them best ; or it may be considered as a laboratory rather than a mine, in which the product is in ratio to the intelligence which is applied to its use.

suffices for the supply of 500 persons. One operator in the cotton factory makes cloth for 250, in the woollen factory for 300; one modern cobbler (who is any thing but a cobbler), working in a boot and shoe factory, furnishes 1,000 men, or more than 1,000 women, with all the boots and shoes they require in a year. So it goes on; and the more effective the capital, the higher the wages, the lower the cost, the more ample the supply.

But in the consideration of this or any other theory of wages, it must always be remembered that these natural laws which govern the actions of men in the conduct of the processes of industry, work very slowly, and are subject to variable causes or interruptions which may suspend, retard, or even reverse their normal action for a considerable period. For instance, the process of making iron, beginning with the mining of the coal and of the ore and ending with the conversion of the materials in the furnace, calls for the use of a very large capital, and for the highest scientific attainments in the heads of departments and in the administration of the work. It also requires special skill on the part of a small portion of the workmen, but the larger part of the work is not of the kind that calls for any great measure of intelligence, and is, in fact, mainly handwork. It might therefore happen that the country which first engaged in this branch of industry on a large scale would obtain a paramount control of all markets and might be able, for a long period, to prevent the building up of competitive works elsewhere. In fact, so long as the only fuel with which iron was smelted was charcoal, the colonies of America were able to supply themselves, and even to export large quantities of iron to Great Britian. But when a method was invented for the application of coal to the smelting of iron, the supremacy of Great Britain in this art was assured for a long period. A dense population gathered round her mines, skilful enough for this work, but otherwise unintelligent, uninstructed, and irremovable, or practically incapable of meeting the conditions necessary for beginning this work in other countries. Under such conditions as these, the British employers of labor in making iron were in a position which enabled them to keep wages down, and to keep prices and profits up for a long period, as in fact they did. Under such relative conditions the competition with all other countries, especially a country like the United States where population was very sparse and capital was very limited, was of necessity long delayed, even though our deposits of iron and coal are so placed as to be more easily worked. And even though a ton of iron made in the United States now represents a much less quantity, or less number of days of labor, than a ton of iron produced in Great Britain, it was not always so. It therefore became a mere question of expediency whether or not to interpose a temporary protective duty in order to overcome certain artificial conditions. It was held that a country should render itself substantially independent of all other countries in the making of iron, because iron is one of the essential articles of war. These arguments were entitled to all the consideration which they may deserve. No opinion need here be expressed upon them.

The same retardation in the working of natural laws also occurred in respect to the inventions of Arkwright and others in cotton-spinning. England succeeded for a long time in retaining control of these inventions, which were of prime importance, by making it a penal offence to carry drawings or models to any other country. By this joint control of the processes of making iron and the application of machinery to the cotton manufacture, England obtained the supreme control for a time of this latter art, and fairly succeeded in preventing these modes of work from being carried to this or any other country for very many years. The cotton manufacture was not established in this country until Samuel Slater succeeded in building machinery from memory, having been unable to bring plans from England; of course such an undertaking was at a great disadvantage. In this case, again, the main question as to the development of textile establishments by means of a protective duty became one of expediency only. The expediency of these protective duties was sustained upon the ground that although the people were for the time subjected to the necessity of paying higher prices for their iron and for their textile fabrics than they would otherwise have paid, an ultimate reduction of cost and of price to a much lower plane was thereby assured, and has doubtless been accomplished.

These two examples are cited in order to show that this theory of wages does not of necessity carry with it the *laisser faire* idea of legislation. It is not denied that special branches of industry may be promoted by legislation of this sort. It is not denied that wages in that special branch may be temporarily raised, because by means of the obstruction to foreign import which the duty interposes, the price of the domestic fabric is for a time maintained at a higher point than it would otherwise be ; and since the sum from which wages and profits are alike derived is the value of the joint product, it follows that, in these particular arts, so long as the protective duty serves to keep up the price, there may be more money to be divided in rates of wages to the operatives who do this special work.

But, it will be observed that such additional profit or additional wage is at the cost of the consumer in the same country, and that there can be no material effect upon the general rate of wages because the number of persons now engaged in any branch of industry which could be subjected to foreign competition is very small in ratio to the whole number of persons engaged in gainful occupation. Such duties may be expedient or not. That is not the question at issue in this treatise. I cite these cases in order that the true theory of wages may not be prejudiced in the mind of any one by any apparent antagonism to the protective theory, which may be justified on entirely independent grounds.

In the judgment of the writer the source of wages and the law by which they are determined fail to be comprehended, both by the advocates of protection and free trade, and this failure leads to much useless and bitter contention. If the honest advocate of protection were once convinced that when an industry had become fairly established the rate of wages determines itself according to the general average of wages in other work of analogous kind, and that the wages thereafter tend to the share of the laborer becoming greater and greater, he would be less averse to considering the date when the protective duty could either be reduced or removed. No one but the most confirmed *doctrinaire* can deny that the argument in respect to wages and to their maintenance which is presented on behalf of a protective tariff, is conscientiously presented in the interest of labor on behalf of those who adhere to it.

On the other hand, if the equally sincere advocate of free trade could once be convinced that the continued imposition of the duty does not of necessity involve the continued taxation of the many for the benefit of the few; if he could admit that it might even be expedient, under certain circumstances, for the State to grant a special privilege to some special branch of work for a certain period of time, much foolish talk, bitter contention, and absurd misrepresentation would be avoided.

The tariff question, the protection of women and children in factories from overwork or from injury, and other like subjects of legislation, are questions of expediency, varying with the time and circumstances of each country. They are not like slavery or inconvertible paper money, moral questions, upon which no compromise can be tolerated; but, on the contrary, they are subjects for reasonable consideration and for reasonable compromise among honest and fair-minded men. When the whole direction of domestic industry has been in some measure altered by the continued imposition of high duties upon foreign imports which were the necessity of war, nothing could be more injudicious than to adopt revolutionary changes. It may have been bad policy to impose the high duties, but it does not follow that it would be good policy to remove them all at once, or that he is a spoliator who asks time to adjust his capital and the labor which he employs to other conditions.

I have recited the various changes which have affected a single textile art. Periods of prosperity and adversity affect all commercial and manufacturing countries alike. They are more intense in one country than another; sometimes most intense in a country which, like Great Britain, depends upon the widest foreign commerce, sometimes in a country which, like the United States, depends mainly upon domestic commerce. Statutes in regard to the collection of revenue, the hours of labor, and the like, may make these fluctuations a little more, perhaps a little less intense, but in the long run they have and can have no permanent effect. Competition adjusts itself to all conditions, and, in the long run, wages or earnings will be the highest in that country in which capital and labor coöperate to the fullest extent, thereby assuring the largest production at the lowest labor cost.

The progress of the United States has been uniformly onward, despite all the vacillations and changes in her financial policy. Our greatest dangers and most serious disasters have arisen from bad money rather than from bad methods of taxation. The danger now before us, growing out of the continued coinage of a silver dollar of light weight, is perhaps the most serious one. Next to that comes the danger growing out of the enormous excess of our national revenue; but even this enormous excess of revenue will itself force upon us a change in our method of taxation. In that again comes a danger, because next to the evil which may be inflicted upon a country by the imposition of heavy taxes, is the evil which may come from an injudicious method in removing them after the industry of the country has adjusted itself to them.

I have endeavored to separate the fundamental principle of wages from all such side issues, and to prove, with as much scientific accuracy as may be possible, that the interests of the employer and the employed are absolutely identical, and that progress and poverty are not of necessity evolved together under the existing customs of the English-speaking people. I have referred to the admirable address of Mr. Robert Giffen, proving a similar progress to that of this country in Great Britain, and from similar data. I had not read that treatise until after the substance of this essay had been compiled.

Let me refer finally and but a moment to one great cause of disturbance in the relations of men to each other. The inventor, the man of science, is the great disturber of existing conditions. He renders worthless great masses of capital which had been valuable; he takes away the hereditary occupation of vast numbers of laborers who may be capable of doing no other kind of work. In the process of adjustment to these new conditions many hardships arise, but the end is progress, both in wealth and in the alleviation of poverty. The only accumulation which has any permanent value consists in that experience and versatility, in that habit and capacity of applying brains and hand alike to any kind of work which is waiting to be done, whereby men are enabled to prosper under any and all conditions. The only capital of any importance, which can be transmitted from one generation to another is this power of applying brain and hand together to useful work, whatever may be the changing conditions under which the work of each generation must be done.

Poverty may for a time ensue, as the consequence of invention and the consequent displacement of labor; but it will be observed that this poverty does not ensue either from the accumulation of capital or from the private ownership of land, so much as it does from the destruction of capital and in taking away the value from land.

The jenny and the mule destroyed the spinning-wheel; the power-loom destroyed the hand-loom; the railroad is destroying the canal; the railroad is reducing the value of land in one place and increasing it in another. The discovery of coal oil would have destroyed the candle market, were it not that a demand for the altars of the Catholics continued to sustain a few candle works. The gas engine is destroying the small stationary steam-engine in England, and will soon do so here. Sir Henry Bessemer has taken from the English landowner all power to collect any rent from land devoted to wheat. With each of these changes the few suffer for a time, but the many gain in welfare With each of these changes the proportion of capital necessary to a given production is decreased ; great fortunes are lost, unless the owners of such fortunes can adapt their machinery to all the changing conditions ; but while some fortunes are thus destroyed, others are gained. At the present time, or we may say for the last three years, half the iron works in the United States have been out of blast, and many will never come into blast again ; but during the same three years the production and consumption of iron has been greater than in any other three years since the continent was settled. True prosperity may be guaged by the consumption of iron in all the arts of life, about as surely as by any statistical method. The loss of fortune to a few producers of iron is of no consequence except to themselves, if more iron be provided for consumption. Most of these changes come gradually ; some of them come suddenly. What are called hard times induce the grestest progress. The great crops in this country increased every year during the war, such was the incentive to invention, which became almost compulsory in consequence of the withdrawal of a million men from productive industry.

I have compared the cotton-mill of 1830 with that of 1883, in the same mill-yard ; but there is little left of the factory, either mill or machinery, of 1830; and if there were it would be almost useless. The saving in the cost of moving merchandise over existing railroads, comparing one year with the next preceding, that is, over the railroads existing in each year, has far more than equalled the cost of building all the new railroads constructed in the subseqent year for fifteen years, from 1865 to 1880. In other words, the reduction in the charge on existing railroads each year, computed on the quantity of merchandise moved in that year, has amounted to a sum equal to the sum expended in the extension of railroads in the next year, for each and every year since 1865.

We have been treating only a question of material welfare: What makes the rate of wages? One answer at least we may surely give. When head and hand are rightly trained together so that a man can do the work which is always waiting to be done, whatever the rate of wages may be, it will suffice for the purchase of good subsistence. He who combines the greatest skill of head and hand in useful work will make that exact progress in the accumulation of wealth which will be the just measure of the services which he renders to his fellow-men. In the last analysis the rate of wages rests wholly on character and capacity and under such conditions, the advancement of science is but another name for progress in human welfare.

I am well aware that there is nothing original in the statement of the fact that the application of machinery to production has a tendency to increase the wages of the workman, and at the same time increase the purchasing power of the money in which wages are paid. This is a truism, but how seldom is it comprehended! Apparently never, in the ordinary discussions. Neither employer nor employé can regulate the rate of wages which is to be paid in money, by any bargain or agreement covering a long period. If one employer agrees to pay a higher rate than his competitors, it will only be a question of time when his business will become unprofitable and he must become bankrupt, unless he uses more effective machinery, and thus assures a larger product from a less number of laborers. If any considerable number of employers secure the work of laborers at a less rate of wages than others in the same kind of occupation, unless there is some compensating advantage to the workman in their special establishments, the mere fact that the laborer is willing to work at such less rate proves him to be incapable or inefficient, and therefore his work will be of high cost.

I have attempted to demonstrate that in all productive employment the rate of wages which can be paid in money must depend on the sum of money which is received from the sale of the product. Inasmuch as those who work for wages in strictly productive occupations constitute by far the largest portion of wage receivers, the rates of wages for personal services, which are only indirectly productive, are gauged by the same standard. All profits and wages must come out of the gross product. Furthermore, all profits, wages, earnings, or other income, must be substantially derived from each year's product, because the year corresponds to the series of seasons in which one crop is made. A part of the product of each year is carried over to start the work of the next year upon; but a part of the product of the present year was brought over from the previous year to start the work of this upon. Therefore the measure of what there is to be divided by the measure of money must, in the long run, depend upon what each year's product will bring in money. If, then, the annual product is large, because the resources are great, because capital is ample, because labor is effective, because the army is but a border police,—then the sum of money derived from the sale will also be large, for the reason that in spite of all natural obstructions between one nation and another, the product of one nation, as a whole, comes directly or indirectly, into competition with the product of the world.

If the propositions submitted in this treatise can be sustained—to wit: that wages are a constantly *increasing* remainder over after *lessening* rates of profit have been set aside from an *increasing* product, it follows that the ability of a very productive country to find a market for its excess, especially of farm products, is a most important factor in determining the price of the whole product, and therefore in determining the general or average rate of wages and profits which can be recovered from the sale of the whole. Hence arises the importance of our foreign export of the products of agriculture. Even though the quantity exported be but a tithe of the whole, yet the sale of this part determines the price of the whole, and it therefore becomes a prime factor in the general rate of wages.

If this latter statement be questioned, it will only need a moment's consideration to determine it. If the surplus or over-production for domestic use, of our oil, grain, cotton, meat, cheese, butter, lard, etc., could not be sold in or exchanged for the products of other countries, what should we do with it? We could not now consume it ourselves; we could not move people from other countries here in sufficient number to consume it in any one year. We cannot establish manufactures more rapidly because goods are already in excess. We must exchange our excess for tea, coffee, sugar, hides, wool, and the like, and in the process of this exchange the price of all our crops is determined by what this excess will bring; the remainder over from these sales establishes the standard of farm wages, by, or in comparison with which, all other wages are in the main determined. Hence the average rate of domestic wages rests in a very great degree, under our present conditions, on our finding a foreign market for the excess of our products of agriculture; if this market is limited or reduced, the purchasing power of our farmers, numbering one half our population, is reduced, and this re-acts on the demand for domestic manufactures. Thus it is, that directly or indirectly the value of our total production is determined by a world-wide competition. What would be the effect of the competition of the laborers who now engage in the production of that which we export if they were forced into other work for domestic use only?

The number of persons engaged in each specific crop is not given separately in the census, and can only be inferred by deducing relative numbers from the proportion which the value of each crop bears to the value of the whole. The total number of farmers and farm laborers listed in the census was 7,670,493. On the bases of relative values, about two and a half per cent., or less than 200,000, were employed in the production of sugar, wool, swamp-rice, hemp, barley, and a few other articles which may be in part imported from foreign countries.

On the other hand, on the maximum estimate of the total value of all the products of agriculture or of the pasture, over seventeen and a half per cent. was the declared value of the export of farm products. From which it may be inferred that over 1,300,000 farmers and farm laborers were employed in meeting this foreign demand.

May it not therefore be said that all commerce, both domestic and foreign, is a process of liquidation, by means of which the respective shares of capital and labor are determined, each becoming a larger share of a larger sum recovered from such sales, the wider the exchange of product for product, and the greater the service which each renders the other, whether capitalist or laborer.

Finally, the rate of wages, measured in terms of money, can only be determined by dividing this remainder over, after capital has received its compensation, among the laborers who do the work; the respective share of each laborer is then rated only by his or her individual skill, industry, and integrity. In the end character and capacity determine the relative rates of wages of those who do the work.

I may conclude by again referring to the proposition of Frederick Bastiat, which is the motto of this essay: All interests are harmonious. "In proportion to the increase of capital the absolute share (of the product) falling to capital is augmented, but the relative share is diminished, while the share of the laborer is increased both absolutely and relatively."

APPENDIX I.

THIS appendix will be very uninteresting except to students. A summary of its contents may, therefore, be given for the benefit of readers who do not care to go over its dry details, as follows:

Approximate estimate of the value of annual product of the census year		$10,000,000,000
Domestic farm consumption estimated		1,000,000,000
Commercial product		9,000,000,000
Estimated profits of capitalists	$450,000,000	
Estimated savings of other classes	450,000,000	900,000,000
Wages fund		8,100,000,000
Number of persons engaged in all gainful occupations in round figures	17,400,000	
Deduct soldiers, marines, and persons engaged in subordinate positions in the Government service	100,000	
Remainder	17,300,000	
Administrative force *i. e.*, mental rather than manual work		1,100,000
Working force, *i. e.*, wage-earners or small farmers		16,200,000
Average remuneration of the administrative force, per year		$1,000
Average wages or earnings of the working force, per year		$432
Gross amount of national, State, and municipal taxes in census year over		$700,000,000

or eight per cent. of the commercial product.

Each worker is one of a group of 2.90 persons; therefore each average person in a workman's family must find shelter, subsistence, clothing, and pay taxes out of what forty to forty-five cents a day will buy.

Each five cents' worth added to each person's share, or each fifteen cents added to each workman's wages per day, implies, at the present time (1884) an additional product and sale of commodities worth one thousand million dollars a year, which is about the present value of our wheat product, of our pig-iron product, and of all our textile fabrics of cotton, wool, and silk combined.

In the text of this treatise I have presented certain estimates of the value of the annual product of the United States in the census year; also estimates of the gross amount of the profits of capital; and, finally, estimates of the gross amount of wages, which, divided by the number of persons engaged in all occupations, yielded certain rates. The treatment of this subject *in extenso* belongs more to the science of statistics than to the science of political economy. For very many years this branch of work has been a subject of very great interest to me, and many years since I analyzed the returns of the Massachusetts census of 1875, which census remains to this day a model of accuracy of its kind.

Upon the basis of the facts developed in that census, I have endeavored to continue the treatment of the subject, and to consider the larger figures of the census of the United States. In all such undertakings, he who accepts the actual figures, without change or alteration, will be sure to be misled. I concur fully with the opinion of other special census experts with whom I have consulted, as to the qualifications which are necessary to be made in making use of many of the tables of the United States census. I cannot give these qualifications in better words than in those of Mr. Joseph D. Weeks, the special expert who investigated the general subject of wages in the manufacturing industries. His views are as follows: "The census year was in many industries a year of remarkable prosperity. The number of persons employed in certain industries at the close of that year was very much in excess of the number of persons employed at the beginning. In most instances the census gave, not the average number of persons employed in a given establishment during the year, but the number of persons employed at the close of the year. Now it will be manifestly unjust to divide the amount of wages received in that industry for the whole year by the number of persons employed at the close of the year, and say that was the average earnings of the workmen engaged in that industry. The wages are for the whole year, and the number of employés very much in excess of the average for the year. We have also found, as the result of experience, that when workmen do not secure work in their own occupation, they go into others, working in many cases for themselves. For example, our coal miners on the Monongahela River have worked on the average only eight or nine months in the year. The idle time is generally in the summer. Many of them own little farms, and during the slack season for coal mining they are engaged in working their farms ; while others, not having farms, seek employment with the farmers of the neighborhood."

In the census figures which I shall adduce, in sustaining the averages of earnings which I have reached by other and very different methods, this qualification will be applied according to my own judgment, or in accordance with the information which I have received from other special experts ; and I think all who are accustomed to make judicious use of statistics will concur in the opinion that approximate accuracy has at least been attained.

For instance, in the production of a little less than 4,000,000 tons of pig-iron in the census year, according to the figures given by Mr. Jas. M. Swank and Prof. R. Pumpelly, two of the most competent special experts, the number of men and boys employed was as follows :

In coal mines producing that part of the coal which was used in iron furnaces, about	20,000
In iron mines	31,668
In blast furnaces	41,875
Total	93,543

The sum of the wages of this force was $28,458,822 or $305 each, on the average. This appears to be an excessively low rate. But there is little doubt that this payment covered the work of substantially nine months only, and in order to reach a true statement of the average wages in the production of pig iron in the census year, we must add about one third, thus giving an average in all the several departments of the work of $400 per year, again sustaining my computation of the general average, which is given hereafter at $400 nett for each person employed in any kind of gainful occupation.

I have assumed in the body of the treatise that $520 represents, on the average, the full measure of all that is produced by each person engaged in gainful occupation in the United States, and which comes into the market for sale or exchange. I have also assumed that ten per cent. of all that is produced may be set aside, in a normal year, for the maintenance and for the increase of capital, but the larger part of this profit is enjoyed by but a small portion of those who do the work. The greater part of the wage-earners save but little. I have assumed an estimate of the value of the annual product as $10,000,000,000. I have set aside one tenth part for the domestic consumption of farmers and their families. In the list of the occupations of the people of the United States, which is probably one of the most accurate of the enumerations, a little less than one half of the number of males employed in any gainful occupation are listed as farmers and farm laborers, numbering 7,670,493 persons out of a total of 17,392,099, but as those who are engaged in agriculture are mostly men, this force probably sustained at least one half the population, or 25,000,000 persons. The estimate of $1,000,000,000, as the domestic consumption of this half of the population, therefore assigns $40 a year to each agricultural person as the value of the product consumed upon the farm, which is not included in any commercial or census estimate of the value of the annual product. The remainder of the annual product is $9,000,000,000 in value by my estimate, which would constitute the annual value of the *commercial* product, or that part of the product which is bought and sold.

The next question is, What part of this remainder accrues to capitalists or to owners of land, in the form of profits, interest, or rent? I have set aside five per cent. upon the annual product which comes into the market,—that is to say, $450,000,000 as the possible share of capitalists. The remainder of the commercial product is $8,550,000,000. I now set aside five per cent. more upon the commercial product, to represent the profits of business and the savings of working people, $450,000,000. Again we have a remainder of $8,100,000,000, which is subject to division in the way of salaries, wages, or the earnings of small farmers.

Before we compute the sub-division of this remainder, it will be necessary to devote a few paragraphs to national wealth, and to the national profits or savings which are possible ; that is, to the increase of the national capital.

I feel less assurance in respect to the estimate of that part of the annual product of the United States which can be set aside for the maintenance and increase of capital than in respect to the general estimate of the portion which goes to those who do the work. I have estimated the savings or addition to capital at $900,000,000 in the census year.

It will be observed that the measure of the savings of the nation is something quite different from the measure of that which would constitute the profits of individuals ; for instance, the manufacturer or merchant may make a very considerable profit out of his work, but he then distributes a very large portion of this profit in his family expenses, thereby sustaining a large number of persons who are included among the so-called working classes or wage earners.

The final end or contribution to the capital of the nation is therefore a very much less sum than the apparent profit which accrues either from the rent of real estate or from the income derived by the individual owners of manufacturing, railroads, or other investments, or from business.

There are very few data available to an individual student whereby even an approximate estimate of the net savings of the nation can be determined.

My deduction from many methods of analysis is that the normal proportion which can be set aside for the maintenance or increase of the capital of the nation can not exceed ten per cent. of its annual production, and is probably less.

It would perhaps be useless to give examples of the various methods by which I have attempted to determine this point : one will suffice.

The officials of the Census Department have made a very careful investigation in respect to the total amount of property assessed for taxes in the United States, and have extended this sum so as to cover the absolute wealth of the country. The total valuation made by the local assessors for purposes of local taxation was as follows, for the year of which a return was made in the census of 1880 :

Value of real estate	$13,036,766,925
Value of personal estate	3,866,226,618
Total	$16,902,993,543

which sum divided by the population gives $337 per capita, but the valuation for purposes of assessment varies greatly in different States, and a very large proportion of actual property is either exempted—such as a large part of the railway system—or else it escapes taxation.

The census valuation of the actual or absolute wealth of the United States is as follows :

	IN MILLIONS.
Farms	$10,197
Residence and business real estate, including water-power . .	9,881
Railroads and equipment	5,536
Telegraphs, shipping, and canals	419
Live stock, whether on or off farms, and farming tools and machinery,	2,406
Household furniture, paintings, books, clothing, jewelry, household supplies of food, fuel, etc.	5,000
Mines (including petroleum wells) and quarries, together with one half the annual product reckoned as the average supply in the hands of the producers or dealers	781
Three quarters of the annual product of agriculture and manufactures, and of the annual importation of foreign goods, assumed to be the average supply in the hands of the producers and dealers . .	6,160

Churches, schools, asylums, public buildings of all kinds, and other real estate exempt from taxation	2,000
Specie	612
Miscellaneous items, including tools of mechanics	650
Total . . . . . ($43,642,000,000) .	$43,642

It will be observed that in this estimate of wealth the value of land is included.

It is computed that four fifths of the valuation of the farms consists of the land, and from one half to two thirds of the estimate of the residence or business real estate also consists in the value of land.

It will also be observed that the estimate includes household furniture, paintings, books, household supplies and the like, as well as churches, schools, asylums, and public buildings, and that the estimate of the value of railroads is taken at the normal amount of stock and bonds issued, the true cost and real value being much less.

If we separate from this estimate that part of the valuation which consists in the mere value of land, and also setting aside churches, asylums, and the like, which represent wealth consumed rather than reproductive capital in the ordinary use of that term, the total amount would be reduced to at most twenty-five thousand millions, and perhaps to a less sum, and this would represent the actual capital or labor saved for purposes of reproduction during the whole period of the existence of the States and colonies of America, thereby sustaining the commonly accepted proposition, that the value of the actual capital of the richest state or nation can bear a ratio to the value of its annual production of only two to threefold.

The invaluable part of the capital of a nation is that portion which has become a part of the *common wealth*, for the use of which no price can be charged,—such as the opening of the common ways, the removal of obstructions to the navigation of waterways, the clearing of arable land, and other results of labor of the same kind ; but yet more potent in reproductive enterprise is the immaterial capital which ensues from our increasing command over the forces of nature, and our power of directing them to the service of man.

It is admitted by all statisticians of repute that *all* valuations of national wealth which are made in terms of money, like the foregoing, must be used with great caution, and are very liable to mislead, especially when made use of to compare one period with another.

Such comparisons, when honestly made, are rather an indication of ignorance or incompetence in the use of statistics, than of any thing else.

For instance, witness the census data :

In 1860 the assessed value of all the property of the U. S. was given as being	$12,034,560,005
True valuation, estimate	16,159,616,065
Excess of so-called true valuation over assessed value	34 per cent.
In 1870 the assessed value was	14,178,986,732
True valuation, estimate	30,068,512,507
Excess of so called true valuation	112 per cent.
In 1880 the assessed value was	16,902,993,543
True valuation, estimated	43,642,000,000
Excess of so-called true valuation	158 per cent.

It is perfectly well known that a great deal of attention was given to the attempt to ascertain the true valuation of 1880, and very little in 1860 ; while the figures of 1870 are vitiated and rendered almost worthless by the depreciation of the currency at that date.

Hence, any one who should attempt to picture the progress of the nation by a statement that we have gained thirty thousand million dollars (!) in wealth in twenty years, or fifteen hundred million dollars (!) a year, would be obliged to defend the honesty of his purpose by an admission of his utter ignorance of the subject.

In the first place, the statements are wholly misleading, because the value of land is included, and therefore the increase in its value forms an element in the case.

Second, unless such increase in the valuation of land, and of capital placed upon the land, has been accompanied by a greater proportionate increase in the annual product of both, out of which the people may be subsisted—then an increase of wealth on the part of the few who own the land would only be evidence of an increase of want on the part of the many who consume its products.

Third, because the data of 1860 were absolutely incomplete and almost worthless.

Such estimates and comparisons of wealth have their use, but their use is only or mainly in their connection with annual production and distribution. It is doubtless true that this country has made greater progress during the last twenty years, both in wealth and in productive capacity, than ever before. The reasons are plain—three of the principal causes may be cited:

1. The abolition of slavery.
2. The application of machinery to agriculture.
3. The extension and unification or consolidation of the railway system.

It may possibly be true that one half the apparent difference in wealth between 1860 and 1880 represents an actual addition to the productive capital of the country. One half would be $1,500,000,000, or $750,000,000 per year. During this period the average population of the country has been 40,000,000 persons, and therefore such a gain would be at the rate of $18.75 to each person in each year.

When viewed in this aspect, the statement in hundreds of millions is reduced to terms of easy comprehension, and the result indicates the very slow rate at which capital can be accumulated and maintained, rather than the reverse. It must also be remembered that whatever the gain in wealth may be, it is enjoyed by a very small portion of the population.

On the other hand, the taxes which have been imposed during this period have been little below $18.75 per head, if we take into view only the actual assessed taxes during and since the war. In the census year, the aggregate of national, State, and municipal taxation was over $700,000,000, or over $14 per capita, and if the war taxes be computed in their ratio to the population of that date, the sum of all the taxes imposed upon the people of this country since 1860 has without question been equal to at least eighty per cent. of the whole sum which has been added to our productive capital during the same period.

In this light the importance of a correct estimate of the value of our annual product, of the possible profit thereon, the method of its distribution, and the incidence of taxation, become apparent. I have made use of the census estimates of national wealth only for the purpose of rendering the importance of this latter investigation more apparent, and not because I attach much value to statements of accumulated wealth when measured in terms of money.

In pursuance of the main subject, it appears that the sum of national taxes which have been imposed by the Government of the United States upon the people during the last twenty years has been over $7,200,000,000.

The amount of State, county, and municipal taxes for the year reported in the census was over $300,000,000, or $6 per capita. This is at a less rate than for a few years preceding, and at a less rate than was imposed during the war and the years immediately subsequent thereto. If this rate of $6 per capita be applied to the average population for twenty years, the gross amount of such taxes has been not less than $4,800,000,000. The total amount of taxes, therefore, including national, State, county, and municipal, in twenty years, has been $12,000,000,000, or at the rate of $600,000,000 per year.

This sum bears the ratio, for the whole period, of *eighty per cent.* to the sum which I have computed as the true addition to the capital of the nation during twenty years, yet, in spite of this burthen, we have prospered, and have gained in general welfare as well as in national wealth.

At the risk of wearying the reader by repetition let me state this in another form, admitting that there has, without question, been an abnormal increase in the capital of the United States since the end of the war, the chief factor of which abnormal increase has been the saving in the cost of moving commodities by railway, can we measure this single force in any way?

In a treatise upon " The Railroad, the Farmer, and the Public," reprinted herewith, I have clearly proved the fact that had the merchandise, one half of which consisted of crude farm products, which was moved in the year 1883, been subjected to the average charge per ton per mile which was charged on the whole railway service of the United States from 1866 to 1869 inclusive, the sum of such charge in 1883 would have been between twelve and fourteen hundred million dollars, in place of an actual charge of five hundred and fifty million dollars. Between these two periods the value in terms of gold of the principal farm products of the United States, which constitute at least one half the substance moved by the railway, has varied in very slight measure; hence it follows that by far the greater part of the actual saving of labor which has been brought about by the extension and effective working of the railway system, had inured to farmers up to that time.

This addition to our wealth has been in very great measure applied to an increase of capital in railroads, to improvements upon farms and farm buildings, and to various arts and manufactures which must of necessity be carried on near to the farmers upon whom they depend for a market.

Again, I have set aside, by the estimate of the census year, nine hundred million dollars, or ten per cent. of the commercial product, as the probable proportion of the annual product which could be applied to the maintenance, improvement, or increase of capital in that year. This was at the rate of $18 per capita of the population of that year.

The population of the United States has averaged forty million for the whole term from 1860 to 1880, or substantially that number. Multiply forty million by $18, and we have the average sum of seven hundred and twenty million dollars each year, corresponding to my estimate of the census year of nine hundred million dollars. Multiply seven hundred and twenty million dollars by twenty years, and we reach the sum of fourteen thousand four hundred million dollars, set aside from the production of the twenty years, for the maintenance and increase of capital. Deduct fourteen thousand four hundred million dollars from twenty-five thousand million dollars, which appears to be the utmost part of the census estimate of total wealth in 1880 which can be considered the work of man, and we leave only ten thousand six hundred million dollars as the saving of the nation through its whole previous history. This may perhaps lead to the conclusion that my estimate of ten per cent. now set aside as capital, is a reasonable or perhaps excessive estimate of that part of the annual product which, in a normal year, can be set aside for its maintenance or increase; $900,000,000 being ten per cent. of an estimated salable product of $9,000,000,000. If, during the last few years, there has been an abnormal increase of capital at the rate of more than $18 per capita, it has not been at the cost of the laborer, but it has been only a small part of that which the capitalists have themselves saved to the people in the extension of the railway system, and in the erection of factories, mills, and works of various kinds of the most productive and effective sort. This abnormal increase of capital has now ceased, and the prices of farm products are now falling.

It is now probable that the great forces which I have recited have in some measure, or for the time, become exhausted, and that the present period of depression indicates a great change or adjustment of prices on a lower plane and of a permanent character, which will be ultimately beneficial, but which in its progress is disastrous to many and very hard to be borne by all; because in such a period constructive enterprises are checked, and the existing population lives from hand to mouth, anxious as to what each day may bring forth.

In such a period excessive taxation becomes an intolerable

burden. This burden is to be measured by the ratio which the sum of all the taxes bears to the possible sum of all the savings of the community, rather than by its ratio to the gross value of all products ; in other words *by its ratio to net income rather than by its ratio to gross income.* It will be borne in mind that, with very few if any exceptions, all taxes are distributed, wherever they may be first imposed, and ultimately fall on all consumers in almost the exact ratio of their consumption.

If imposed upon dwellings, they are charged to occupants with their rent, or their rent is enhanced so as to cover them.

If imposed upon machinery or other instrumentalities of production, they are charged to the cost of goods and are recovered from the sales.

If imposed upon railroads, warehouses, shops, or other instrumentalities of distribution, they are charged to the cost of distributing goods.

If imposed upon the goods or wares themselves, whether under a tariff or an excise, they are added to the price and recovered from the sales.

Taxation falls on rich and poor according to their consumption, while profits or savings are sorted under a very different law ; hence even the ratio of gross taxation to the net savings of the nation gives no true measure of its burden, but only brings its weight into prominence.

To the rich a tax constitutes more of an annoyance than a heavy burden ; to the man of moderate income it merely causes a slight decrease of comfort or a small reduction in savings ; from the skilled workman it may take half of what he might have saved ; from the laborer it takes even the small pittance that might have served to mitigate the poverty of his later years ; and from the poor it takes a part of what is necessary to existence and reduces them to pauperism. No class of men have so grave an interest in an honest and economical government and in the reduction of taxation than those who possess no property of their own, but who depend wholly upon their daily work for their daily bread.

It is for these reasons that while we may rejoice in the prosperity which has enabled us to reduce our national debt and to put it in the way of final payment within the present century, we may now protest against the excess of taxation which finds men poor, keeps them poor, and will leave them poor, unless it is removed.

Therefore the great issues of the hour are measures not men, and whatever may be the result of the elections now pending, every man chosen will be held to a stern account, and no glittering generalities about the increase of national wealth will serve to meet the demand for relief from the intolerable burden of excessive taxation (Novr., 1884).

Having thus treated the probable profits or savings in the census year, and assuming that my estimates are approximately correct, and that there remained in the census year $8,100,000,000 worth of product to be divided in terms of money between the mental and manual workers, or between the administative and the executive force, in the form of salaries, wages, or earnings, the next problem is the subdivision of this sum. We can reach a close estimate of the mode of this subdivision by a consideration of the details of the census in respect to the occupations of the people and the ascertained rates of wages in special classes ; qualifying the figures by such additions to the rates given in the census as may be called for in each case, as before stated.

We find in the list of all persons engaged in gainful occupations, 1,100,000 persons, under the following classification :

Clergymen	64,698
Lawyers	64,137
Physicians and Surgeons	85,671
Teachers and Scientific Persons	227,710
Actors	4,812
Architects	3,375
Artists, or Teachers of Art	9,104
Authors, Lecturers, and Literary Persons	1,131
Chemists, Assayers, and Metallurgists	1,969
Dentists	12,314
Railroad Builders and Contractors	1,206
Civil Engineers	8,261

Officials of Railroad Companies 2,069
Traders and Dealers 481,450
Bankers and Money Brokers 15,180
Officials of Banks 4,421
Officials of Insurance Companies 1,774
Manufacturers and Officials in Mf'g Cos. 52,217
Hotel Keepers 32,453
Journalists 12,308

Total 1,086,260

This classification is only fairly accurate. If it were possible to get the number, superintendents and foremen should be substituted for about two thirds of the teachers who are in the lower grades.

This class of persons represents those whose work is more mental than manual, more administrative than executive. In round numbers they amount to 1,100,000. The remainder of those who are listed as being engaged in gainful occupations constitute the actual working force—mechanics, artisans, clerks, factory operatives, small farmers and farm laborers, domestic servants, common laborers, express men, conductors, and all others, whose work possesses a commercial value, and whose rate of wages constitutes the measure of their share of the annual product.

Now, then, the last remainder of the assumed annual product amounted to $8,100,000,000. The total number of the actual working force in the list, aside from the administrative force, and recited as above in the census year, was 16,200,000. If to each one of these be assigned a rate of wages upon the average of $432—being the sum which when subjected to the average per cent. or rate of national, State, and municipal taxation, would leave $400 net each per year,—the sum of all their wages would amount to $6,998,400,000. There would then remain $1,101,600,000 to be divided among the 1,100,000 persons of the first class, to wit : those engaged in the mental work, or in the work of administration ; and this sum would yield to each one of these annually $1,000. It will be observed that these conclusions were reached *a priori*, before any consideration or attention had been given to actual rates of wages as disclosed in the census being deduced from an estimate of the annual product reached in the manner previously described.

Before testing these results by the actual data of the census, the total of persons occupied should be considered. It is as follows:

Agriculture, males	7,075,983
" females	594,510
Professional and Personal Service, males	2,712,943
" " " " females	1,361,295
Trade and Transportation, males	1,750,892
" " " females	59,364
Manufacturing, Mechanical, and Mining, males	3,205,124
" " " " females	631,988
Total of all classes	17,392,099
Total, aside from Agriculture	9,721,606
Deduct Civil and Military Employés of the Government in subordinate or minor positions, say	92,000
Total, in round figures, of all persons engaged in any gainful productive occupation	17,300,000
Deduct administrative and mental work	1,100,000
Total in the actual work of production or distribution, who are substantially the wage-earners	16,200,000

The first test by which the approximate accuracy of this estimate of about $432 average earnings may in some measure be determined will be found in the exhaustive treatise of the census, upon Transportation, compiled by the special agent, Mr. A. E. Shuman. This compilation is based upon the actual returns from existing railroads, for specific periods of twelve months, corresponding to the making up of their accounts in the year immediately preceding the census. Now, it is well known that the accounts of railroad operations are of necessity kept in the most accurate manner. Hence these returns may be considered as more closely approximating the actual earnings of the employés than any other returns of the census. It will also be observed that railroad employés are almost wholly men, and that among these men are represented the highest-paid officials, and also the lowest-paid laborers. They number as follows:

General officers	3,375
Clerks	8,655
Station men	63,380
Engineers	18,977
Conductors	12,419
Other train men	48,254
Machinists	22,766
Carpenters	23,202
Other shop men	43,746
Track men	122,489
All other employés	51,694
Total	418,957

The clerks being counted in the work of administration, and the large proportion of well paid engineers and conductors carrying up the executive average of earnings.

The sum of their earnings was $195,350,013, averaging to each person for the year $466. But, upon a further analysis, it appeared that the average earnings of officers and clerks—three per cent. of the total number—amounted to $1,015.44 each; the average wages of all the others—ninety-seven per cent. of the total number—were $450 each. If I am right in assuming that these railroad employés are a fairly representative class of all men employed in gainful occupations, bearing in mind the less rate of earnings of women, these figures, both of the higher grade of administrative work and the lower grade of executive work, fairly correspond to the averages of my assumed figures covering all persons occupied within the limits of the country.

We will next consider another class of persons, chiefly men and boys, to wit: all who are listed as being employed in mining the non-precious metals—iron, copper, lead, and zinc. In a special report upon this industry, it appears that the total number returned is 220,475; the sum of salaries and wages paid, $71,-992,502; an average to each person of $327. But all the census experts concur in the opinion that this sum did not represent over three fourths of a full year. Many new mines were opened during the census year, of which the returns covered only a part of the year; and, as has been stated by Mr. Weeks, work is not continuous, even in mines at regular occupation. If, then, we increase the sum of $327 by the addition of one third, thereby converting the term into a full year's payment, provided these men find employment in other occupations, we reach an average of $436 as the income of each person employed in this arduous work. The proportion of those who are engaged in the work of administration being less than in the railroads, a fair approximation is made to the income of the wage-earners of $400 per year (see previous figures on iron).

The next mode of comparison may be with the average earnings of all persons who are listed in the census under the head of manufacturing. That comprised 2,019,135 men, 531,639 women, and 181,921 children,—a total working force of 2,732,695. The sum of their earnings or wages was $947,953,795,—giving an average to each person of $346. But this result again must be subjected to very important qualifications. The list of occupations listed under the term of manufactures includes brick-making, which can only be followed six months in the year; lumber-men's work, generally limited to six months in the year. Other branches of industry, which are continuous, are again subjected to the qualification named by Mr. Weeks. The writer was one of the special experts employed in taking the census of the cotton manufacture. He began among the first, and gave a construction to the directions which he received, which led him to omit from the number of persons and sum of wages in the cotton manufacture those who were engaged as agents or superintendents in charge of the work. In all other branches of the census he has been informed that the administrative force was included. The wages in the cotton manufacture appear to be only $245 each per year, by far the larger portion of those employed being women and children; but in his judgment this sum should be raised to at least $280 and, including administrative force, perhaps to $300 per year, in order that it may be made to correspond to full year's work of those who were continuously employed.

In general, it may be said that the necessary qualifications by which the average wages disclosed by the census, in respect to all manufactures, should be governed, would lead to the conclusion that $346 represented not over ten months' work. And if then we add one fifth of $346, to make up for the two months, we reach a general average, including the administrative force, of $415 each, —again substantially corresponding with the conclusions of the writer, and again substantially corresponding to the railway figures giving due consideration to the lower wages of women and children.

Subject to these qualifications, the following specific data from the census are given, in respect to branches of manufacture which may be considered substantially continuous. Each branch may be qualified, according to the judgment or knowledge of the reader. It should be noticed that those who are listed under the head of carpentry are only the carpenters who are engaged in manufacturing establishments of which the product exceeded $500 a year ; and it does not include miscellaneous carpenters, who are much more numerous. In all the textile arts the figures should probably be raised at least one fourth, in others more or less according to the special conditions of each case.

	Men.	Women.	Children.	Total.	Wages.	Avge.
Agricultural Implements,	38,313	73	1,194	39,580	$15,359,610	$388
Book Binding and Blank-Book Making,	5,127	4,831	654	10,612	3,927,349	371
Boots and Shoes,	104,021	25,946	3,852	133,819	50,995,144	381
Bread and Bakery Products,	18,925	2,210	1,353	22,488	9,411,328	419
Carpentry,	53,547	77	517	54,138	24,562,077	454
Cars,—Railroad and Street,	13,885	13	334	14,232	5,507,753	388
Carriages and Wagons,	43,630	273	1,491	45,394	18,988,615	400
Men's Clothing,	77,255	80,994	2,504	160,753	45,940,353	286
Foundries and Machine Shops,	140,459	675	4,217	145,351	65,982,133	454
Furniture,	45,180	917	2,620	48,717	20,388,794	418
Jewelry,	10,050	1,998	649	12,697	6,441,688	507
Leather Currying,	10,808	77	168	11,053	4,845,413	438
Leather Tanning,	23,287	188	337	23,812	9,204,243	387
Malt Liquors,	27,001	29	190	26,220	12,198,053	468

	Men.	Women.	Children.	Total.	Wages.	Avge.
Marble and Stone,	21,112	23	336	21,471	10,238,885	$477
Paper,	16,133	7,640	649	24,422	8,525,355	349
Printing and Publishing,	45,880	6,759	5,839	58,478	30,531,627	522
Tobacco, Cigars, and Cigarettes,	40,099	9,108	4,090	53,297	18,464,562	347
Hardware,	14,481	814	1,506	16,801	6,846,913	407
Cotton Goods,	64,107	91,148	30,217	185,472	45,014,419	245
Cutlery and Edge Tools,	9,458	380	681	10,519	4,447,349	422
Glass,	17,778	741	5,658	24,177	9,144,100	379
Hats and Caps,	11,373	5,337	530	17,240	6,635,522	385
Hosiery and Knit Goods,	7,517	17,707	3,661	28,885	6,701,475	232
Mixed Textiles,	17,471	20,520	5,382	43,373	13,316,753	308
Musical Intruments.	6,449	57	69	6,575	4,603,193	692
Woollen Goods,	46,978	29,372	10,154	86,504	25,836,292	300

No data exist by which the earnings of agricultural laborers can be positively converted into terms of money, owing to the fact that by far the larger portion receive a part of their wages in kind, and not in money. By the courtesy of Mr. J. R. Dodge, the statistician of the Agricultural Department, I am enabled to submit the following table of wages of agricultural laborers in the year 1882. Due consideration being given to the domestic consumption of the farmer, I think they substantially sustain my assumed average of the subdivision of the annual product.

Many of these men are engaged in the winter as lumbermen or other occupations, or as stated by Mr. Weeks, in mining, they making up their rate to the full average for the year.

No census data exist by means of which the average earnings of persons engaged in trade or commerce can be estimated. The average of those who are engaged in other kinds of transportation than by rail, to wit, upon rivers, expressmen, and wagoners, may be considered in the ratio which these occupations bear to the railway service. The men who are employed in these other branches of transportation are continually changing, sometimes being engaged upon the railway, sometimes in the other branches of the work.

The average earnings of persons in domestic service can only be established by their known ratio to the work of the factory operative, or of other persons engaged in analogous employments.

FARM WAGES IN 1882.[1]

By the Month, and by the Day, in Harvest; with payment in Cash, and also in Money supplemented by Board.

States and Territories.	Monthly Wages By the Year.		Transient Wages During Harvest, per day.	
	Without Board.	With Board.	Without Board.	With Board.
Maine	$24.75	$16.75	$1.52	$1.22
New Hampshire	25.25	16.72	1.71	1.35
Vermont	23.37	16.00	1.75	1.35
Massachusetts	30.66	18.25	1.75	1.35
Rhode Island	27.75	17.00	1.60	1.30
Connecticut	27.90	17.37	1.65	1.33
New York	23.63	15.36	1.89	1.47
New Jersey	24.25	14.20	2.09	1.74
Pennsylvania	22.88	14.21	1.73	1.30
Delaware	18.20	12.50	1.60	1.25
Maryland	16.34	9.89	1.52	1.15
Virginia	13.96	9.17	1.27	.99
North Carolina	12.86	8.80	1.20	.85
South Carolina	12.10	8.10	1.08	.78
Georgia	12.86	8.70	1.10	.80
Florida	16.64	10.20	1.12	.80
Alabama	13.15	9.09	1.05	.80
Mississippi	15.10	10.09	1.23	.95
Louisiana	18.20	12.69	1.10	.85
Texas	20.20	14.03	1.39	1.08
Arkansas	18.50	12.25	1.34	1.02
Tennessee	13.75	9.49	1.30	1.00
West Virginia	19.16	12.46	1.30	1.00
Kentucky	18.20	11.75	1.54	1.18
Ohio	24.55	16.30	1.79	1.41
Michigan	25.76	17.27	2.13	1.76
Indiana	23.14	15.65	1.89	1.58
Illinois	23.91	17.14	1.91	1.54
Wisconsin	26.21	17.90	2.50	2.10
Minnesota	26.36	17.75	2.61	2.16
Iowa	26.21	17.95	2.25	1.81
Missouri	22.39	13.95	1.59	1.23
Kansas	23.85	15.87	1.70	1.35
Nebraska	24.45	16.20	1.95	1.57
California	38.25	23.45	2.30	1.86
Oregon	33.50	24.75	1.92	1.50
Colorado	36.50	27.08	2.21	1.80

J. R. DODGE, Statistician.

[1] It will be observed that the foregoing list only covers the rates of wages of

The average pay of common laborers in the census year varied from $1 to $1.50 for the working days of the year; but it is well known that the daily rate cannot be considered as a continuous rate throughout the year. The average earnings of common laborers could not have been more than $400 a year; but it may perhaps be admitted that they fairly approximated that sum, again sustaining my assumed figures.

It therefore follows that if the value of the annual product ap-

---

farm laborers. The larger part of the whole number of persons listed as being engaged in agriculture are listed as farmers, and not farm laborers.

The total number employed in agriculture is:

Male	7,075,983
Female	594,510
Total	7,670,493

It is probable that each one of these persons stands at the head of a somewhat larger group than the average group of 2.90 in all arts, and that not less than one half the population, or 25,000,000 persons, were wholly dependent upon this agricultural portion of the working force in the census year.

The primary value of the farm product of 1879 (subject to moderate increase in 1880), as given in the census, is $2,212,540,927, but the census experts point out the necessity of adding materially to this sum, to cover the home consumption of the farms.

I have ventured to add $1,000,000,000 to this computation of the primary value, in order to cover the domestic consumption of the agricultural population, which never appears in the commercial tables, but which should be computed and added to the agricultural product, as well as other almost necessary omissions in the census which should be added in order to show the relation which the work of each person devoted to agriculture bears to the work of each person engaged in other branches of industry.

This was also an *a priori* conclusion, but if we add to the census valuation $2,212,540,927, the sum of $1,000,000,000, for domestic consumption, and divide by the number of persons occupied, 7,670,493, we get an average product of a fraction less than $419 each, again fairly corresponding to my assumed average.

It will be observed that the total number of farms was 4,008,907, averaging substantially seventy acres of improved land each. There were substantially one farmer and one laborer to each farm, and it therefore appears that the average farmer can be assumed to earn but a moderate sum above that of the farm laborer.

proximated $10,000,000,000, the average wages of earning people must have approximated $432 a year ; and upon what this sum would purchase nearly three (2.90) persons were on the average sustained. This gives $147 per year, or 40 cents a day to each person. That is to say, each person on the average was subsisted, sheltered, and clothed on what 40 cents a day would buy from that part of the commercial product available for wages.

If such was the measure in money of all that was produced, which could be made subject to division or commercial distribution, then it will be apparent that there could be no greater sum or money's worth to be divided. If any less part of the product had been set aside for profits or increase of capital than that which I have assigned hitherto, then the increase of capital would have been checked, and the production of the next and of ensuing years, in ratio to the number of existing persons, would have suffered.

There can be no general rise in the rates of wages, except by means of an increase in the quantity of things produced, coupled with the maintenance of the prices at which such products can be sold. There may be an increase in the general welfare, by way of an increase in the quantity of things produced, coupled with a decrease in price, which shall not affect the gross value of the whole, so that the rate of wages may buy more commodities ; or, in other words, may represent a larger quantity of things. There may be increase in the general welfare brought about by the increase in quantity and decrease in price, coupled with a decrease in the money rate of wages, if such a decrease in the rate of wages does not go below the decrease in prices.

I have before referred to the burthen of excessive taxation, but this point cannot be too often pressed. So far as the proceeds of taxation are expended for just administration, for a good government, wisely and honestly administered ; or in municipal affairs, so far as the avails of taxation are expended in the maintenance of good highways, of sewers, in providing an adequate water supply, and in sustaining the common schools,—

taxation cannot be considered a burden, but is a distribution of a part of the annual product, for the common welfare and for the general benefit. But so far as the proceeds of taxation are wasted or misspent, then taxation becomes an intolerable burden, and it must be gauged, not by its ratio to the gross product of the country, but by its ratio to the net income, or to the possible savings of each person. The conviction of the writer is that all taxation ultimately falls upon consumers, in the ratio of their consumption, no matter where the taxation is first laid, whether it be a direct tax upon real estate, or an indirect tax upon certain specified articles. If a tax is laid upon real estate occupied for commercial purposes, it becomes a charge upon the distribution of the goods. If it is levied upon land used for agricultural purposes, it enters into the money cost of production. If it be admitted that taxes are borne in the ratio of consumption, and that producers are merely the agents for their collection, even very heavy taxes may constitute no real burden upon persons who are in the possession of large property or large incomes. They may be but a light burden upon persons of moderate means or moderate income; but when they either restrict the consumption of the necessaries of life, or take from working people the little margin which might be saved, they become intolerable,—if they are either unjust or unnecessary.

If the estimate of the salable or exchangeable value of our annual product which I have assumed in this treatise is even approximately correct, then eight per cent. upon such exchangeable value, aggregating $9,000,000,000, is distributed by way of taxation,—the aggregate of the National, State, County, City, and Town taxes in the census year having exceeded $700,000,000

If I have set aside a sufficient sum to represent profits, to wit: ten per cent. of the total product, half being assigned as the profits of capital, and half being assigned as the savings of those who perform the work of distribution or production, $900,000,000 in all, then the taxes of 1880 bore the ratio of eighty per cent. to the probable savings of the country. If it may have been possible in some one extremely prosperous year since the war, to set aside fifteen per cent., or $1,450,000,000, still the actual taxes bore the ratio to this sum of fifty per cent. Now, if it be true that nine tenths or more of all who are engaged in gainful occupation must subsist, save, and pay taxes out of an average income of $400 to $500 a year, and if of this sum $32 to $40 must be set aside to meet the heavy taxation of this country, it follows that such a burden may not only deprive a very large portion of the working people of this land of the opportunity to save any thing, but may even take from very many of them a part of that which is necessary even for a comfortable subsistence. It follows that the man upon whom the burden of taxation falls heaviest is he who possesses no property whatever. It finds him poor, it keeps him poor, and it may even reduce him to pauperism; yet he may never know the cause of his poverty, and may resist the very changes in the system of taxation which would benefit him most. The writer is of the profound conviction that whenever the subject of taxation is reduced to a science, taxation on real estate will become the source of nearly all taxes. A tax on real estate cannot be evaded; it diffuses itself with unerring certainty; it forces unoccupied land into productive use; it compels the most conservative class in the community to take an active part in true politics, and to watch the expenditures of the Government, whether national, State, or municipal, with the closest scrutiny. Such a tax may perhaps be supplemented by taxation on railways, gas companies or other franchises which are somewhat restricted in their nature and by an excise on spirits collected from the producer; but this opens a broad subject outside the scope of this treatise.

I am aware that some observers compute the value of our annual product at a larger sum than I do, but on the basis of the population of 1880 and the data of that year, I can find no trace of larger earnings or greater profits than my computation would have yielded.

No one can be more aware than the writer of the huge difficulties which occur in computing the accumulated wealth of the country or the value of the annual product, in terms of money. It can only be by bringing these vast aggregates to individual units that an estimate can be made with even approximate accuracy. Attention has often been called in the treatises upon political economy to the small proportion which the aggregate value of accumulated wealth necessarily bears to the money value of the annual product. Owing to the method of taxation, to the various official returns of the States and cities, and to the great skill of Mr. Carroll D. Wright, by whom the census of 1875 was taken, the actual money value both of land and of the capital which has been placed upon the land in the State of Massachusetts can be ascertained with almost absolute certainty. So, also, the value of the annual product of Massachusetts can be approximated with almost absolute certainty. By these figures, it appeared that the absolute value of all the capital of the State of Massachusetts in 1875, *i. e.*, of the mills, workshops, railroads, dwellings, goods, and wares, which had been converted into form for human use by human work, did not exceed three years' annual production. If the data of the census of the United States could be treated in the same exhaustive way, and the value of the land could be deducted from the gross sum of $44,000,000,000, given as the estimate of wealth, it would without doubt appear that the actual capital of the country could not exceed twice or twice and a half the value of its annual product. When the complaint is made that a good subsistence and an adequate shelter can barely be obtained by each three persons upon an average income of only $400 to $500 a year, at the retail value of all they consume of their own production, or procure by purchase or exchange for the three, the only remedy which can be provided is to increase the product. If such is the present measure of all there is, then such is the measure of the utmost that all can have. How difficult and how slow such an increase must be, may be comprehended by a very simple statement : Assuming the maximum of $10,000,000,000 given in this treatise as the present value in the census year, or about 11,500,000,000—now then over $1,000,000,000 worth of produce must be added in a year and the prices must be maintained where they are, in order that each person of our present population may have five cents a day more than they now do, or in order that each person engaged in any kind of gainful occupation may be able to obtain an increase in the rate of wages of fifteen cents a day. Upon such small fractions must subsistence depend, and when political leaders present magnificent pictures of national progress, summed up in thousands of millions of wealth or product, these facts may well be recalled.

Even if our progress has been great and our conditions are relatively prosperous compared to other nations, yet the average person, including capitalists, landowners, employers and employed must have been sustained and sheltered, must have paid taxes and saved profits, out of what fifty cents a day would buy in the census year, *because such was apparently the measure of all there was produced which could be bought and sold or exchanged.*

APPROXIMATE SUMMARY.

Total product of the U. S. $10,000,000,000, worth per day
to each person as estimated 55
Domestic production, consumed without purchase or sale . 5
　　　　　　　　　　　　　　　　　　　　　　　　　　— 50 cts.

Share of capitalists 2½
Savings of the people 2½
National, State, and Municipal taxes 3½
Cost of mental or administrative work 1½
Average to each wage earner 40
　　　　　　　　　　　　　　　　　　　　　　— 50 cts.

For each error of five cents a day in this estimate,—if the reader finds one or believes that there may be an underestimate—add one thousand and fifty-eight million five hundred thousand dollars to my gross estimate and divide the proceeds among the 58,000,000 persons who will probably constitute our population on the 1st Jan., 1885.

APPENDIX II.

THE LAW OF COMPETITION: IN ANY GIVEN PRODUCT, PROFITS DIMINISH, WAGES INCREASE.

The following deductions have been made from the accounts of two New England cotton factories, both constructed prior to 1830, and operated successfully and profitably since that date, mainly on standard sheetings and shirtings—No. 14 yarn. The figures given, from 1840 to 1883 inclusive, are absolute, being taken from the official accounts of mills, of which the sole product has been a 36-inch standard sheeting. The figures of 1830 are deduced from a comparison of the data of two mills. The figures of 1884 are deduced from nine months' work in 1883-4.

WAGES PER OPERATIVE PER YEAR.

Year	Wages
1830	164. gold.
1840	175. gold.
1850	190. gold.
1860	197. gold.
1870	275. cur.
1870	240. gold.
1880	259. gold.
1883	287. gold.
1884	290. gold.

PROFIT PER YARD NECESSARY TO BE SET ASIDE IN ORDER TO PAY 10 PER CENT. ON CAPITAL USED.

Year	Profit
1830	2.400. gold.
1840	1.181 gold.
1850	1.110 gold.
1860	.693 gold.
1870	.760 cur.
1870	.660 gold.
1880	.481 gold.
1883	.434 gold.
1884	.408 gold.

YARDS PER OPERATIVE PER YEAR.

Year	Yards
1830	4,321
1840	9,607
1850	12,164
1860	21,760
1870	19,293
1880	28,000
1883	26,641
1884	28,032

} Changes in the machinery affected production.

COST OF LABOR PER YARD.

Year	Cost
1830	1.900 gold.
1840	1.832 gold.
1850	1.556 gold.
1860	.905 gold.
1870	1.425 cur.
1870	1.240 gold.
1880	.930 gold.
1883	1.080 gold.
1884	1.070 gold.

COMPARISON OF 1840 WITH 1883-4.

This comparison will not show the full reduction in the cost of labor per yard which may be expected in 1884-5, because changes have been in progress which, when completed, will increase the capacity of the mill about 15 per cent., and it is a well-understood rule that, while such changes are being made, the current work of production is done at a disadvantage.

1840-1884.

I.—Capital . . .	1840	$600,000	} Same.
	1883	$600,000	
II.—Fixed capital .	1840	$310,000	} Same.
	1883	$310,000	
III.—Active capital.	1840	$290,000	} Same.
	1883	$290,000	
IV.—Spindles . . .	1840	12,500	} Increase, 146 per cent.
	1883	30,824	
V.—Looms . . .	1840	425	} Increase, 135 per cent.
	1883	1,000	
VI.—Fixed capital per spindle .	1840	$23.20	} Decrease, 57 per cent.
	1883	$10.06	
VII.—No. of operatives emp. .	1840	530	} Same.
	1883	527	
VIII.—Operatives per 1,000 spindles	1840	42 4-10	} Decrease 60 per cent
	1883	17 20-100	

IX.—Lbs. per spindle per day	1840	0.456		Increase,
	1883	0.556		22 per cent
X.—Lbs. per operative per day	1840	10 76-100		Increase,
	1883	31 20-100		190 per cent.
XI.—Hours work per day	1840	+13		Decrease,
	1883	11		15 per cent
XII.—Lbs. per operative per hour	1840	0.83		Increase,
	1883	2.83		240 per cent.
XIII.—Wages per operative pr. y'r	1840	$175		Increase,
	1883	$287		64 per cent.
XIV.—Wages per operative pr. h'r	1840	4.49 cts.		Increase,
	1883	8.80 cts.		96 per cent.
XV.—Wages per y'd	1840	1.82 cts.		Decrease
	1883	1.08 cts.		41 per cent
XVI.—Profit per y'd 10 per ct. on capital	1840	1.18 cts.		Decrease,
	1883	0.43 cts.		63 per cent.
XVII.—Price of goods cost cotton same	1840	9.04 cts.		Decrease,
	1883	7.04 cts.		22 per cent

COMPARISON OF 1830 WITH 1884.

In this comparison the statements are based in part upon the figures of each mill. Both appear to have cost about $40 per spindle, including dwellings for operatives. More than one kind of goods were made in each for a time, but the figures have been adjusted to standard sheetings, an average having been computed by the yard and pound.

Fixed capital	1830	$332,000		Decrease,
	1884	$310,000		37 per cent.
Spindles	1830	8,192		Increase,
	1884	30,824		276 per cent.
Fixed capital per spindle	1830	$40.50		Decrease,
	1884	$10.07		75 per cent.
Operatives per 1,000 spindles	1830	49		Decrease,
	1884	17 2-10		64 per cent.
Pounds per operative per day	1830	9.94		Increase,
	1884	31.22		214 per cent.

The hours of labor in most of the factories in 1830 were 14 per day.

Wages per operative per year	1830	$164		Increase,
	1884	$290		77 per cent.

The wages per hour in 1884 are more than double those of 1830.

Wages per yard	1830	1.90 cts.		Decrease,
	1884	1.07 cts.		44 per cent.
Profit per yard at 10 per cent. on capital	1830	2.40 cts.		Decrease,
	1884	.41 cts.		83 per cent.

In the mountain section of the southern United States the people are still clad in homespun fabrics. Five women—two carders, two spinsters, and one weaver—can produce eight yards per day.

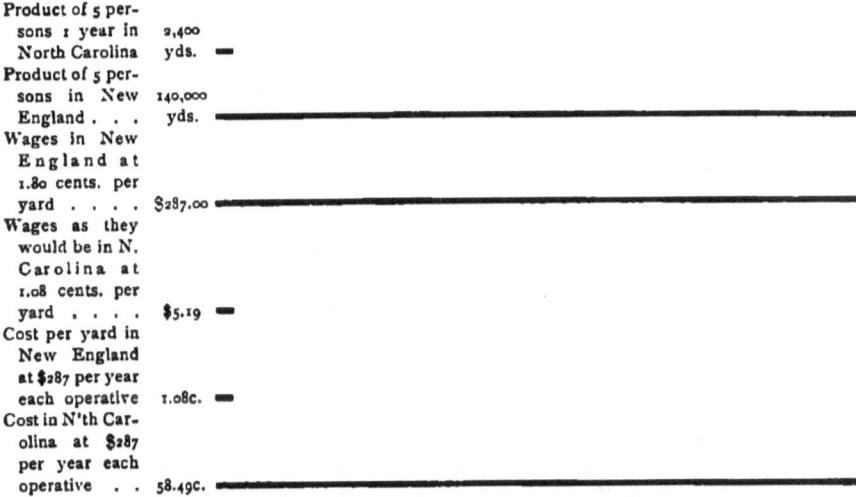

Product of 5 persons 1 year in North Carolina	2,400 yds.
Product of 5 persons in New England	140,000 yds.
Wages in New England at 1.80 cents. per yard	$287.00
Wages as they would be in N. Carolina at 1.08 cents. per yard	$5.19
Cost per yard in New England at $287 per year each operative	1.08c.
Cost in N'th Carolina at $287 per year each operative	58.49c.

The rule of diminishing rates of profit and increasing rate of wages, of necessity ensuing from the progress of invention, is fully sustained by these tables. As the capital is increased both in its quantity and in its effectiveness, the absolute share of product falling to capital is increased, but the relative share is diminished. On the other hand, the share of the laborer is increased, both absolutely and relatively. Labor takes of necessity a constantly increasing proportion of an increasing product. In this example, the wages of the operatives have increased, since 1840, 64 per cent. per day and 96 per cent. per hour; since 1880, 77 per cent. per day and + 100 per cent. per hour. High wages in money have ensued as the necessary result of the low cost of labor.

It will be observed that in 1840 the price of standard sheetings being 9 cents a yard it required 1.18 cents to be set aside for profits, or 13 per cent. of the price, in order to pay 10 per cent. upon the capital. Next it required 1.83 cents to be set aside, being 20 per cent. of the whole price, to pay wages at the average rate of only $175 a year to each operative. In 1884, the price being 7 cents a yard, it required less than 6 per cent. of the gross sales, 0.40 cent a yard, to be set aside in order to pay 10 per cent. upon the capital, while 1.07 cents being set aside as the share of labor, or a fraction over 15 per cent. of the gross sales, yielded to the operative $290 in gold. The goods cannot now be sold at 7 cents, and there is little or no profit for the time being. But while 10 per cent. was a moderate rate of profit in 1840 it is an excessive rate in 1884. The business would extend with great rapidity if there were a positive assurance of 6 per cent. upon the capital, or a quarter of a cent a yard and less than 4½ per cent. of the gross amount of sales.

But it may be said, having assigned 0.40 cent to profits, and 1.07 cents to labor out of 7 cents a yard gross value, there remain 5.53 cents a yard to be accounted for. This of course represents the money cost of cotton, fuel, starch, oil, supplies, taxes, cost of administration, transportation of the goods to market, and the cost of selling them at wholesale.

Does this all go to labor, or is there also a profit to be set aside on these elements?

Our space would not suffice to treat each one of these subjects, but it may be said: First, the cotton is substantially all labor; there is no large margin of profit at the present time in raising cotton, which is mostly produced by small farmers. Second, the other items constituting the materials, form a very small part of the total cost, and are subjected to profits in small measure only in respect to fuel and oil.

The cost of transportation yields to the railroads less than an average of 5 per cent. on the capital invested, and cotton fabrics pay but a small fraction of their value even for very long distances. The cost of administration constitutes a very small part of the cost of the goods, and in a general treatise on wages belongs in a class by itself rather than to be considered as profits. The charge for selling staple plain cotton goods at wholesale does not exceed 1 per cent. to 1¼ per cent., and a large part of this is distributed among the clerks and salesmen who do the work.

If the subject is analyzed, first, as a whole, and, second, in each department, it will appear that at the present time the proportion of profit which can be set aside from the sale of coarse cotton goods sufficient to cover profits in all the various departments of the work, is less than 10 per cent. of the wholesale market value of the product, and 90 per cent. is the absolute share of the laborers who do the work both in respect to materials used and to the finished product.

It is also necessary to remember, in respect to the cotton factory, that the value or proportion of capital to a given product is greater than in almost any other branch of industry; the proportion of capital to product being $1 of capital to each $1 or $1.50 of product, according to the weight of the fabric and the quantity of cotton used. In the boot and shoe factory, on the other hand, the ratio of capital to product is about $1 to $3; therefore in the boot and shoe business a much less proportion of the gross sales needs to be set aside as profit on the business, to induce its being established.

THE RATE OF WAGES? 123

On the whole, so far as the manufactures of New England are concerned, the average of capital to the gross value of the products is one dollar capital to two dollars product; therefore three per cent. of the gross sales set aside as profit will yield six per cent. per annum upon the capital invested in the buildings and machinery which are applied to the conversion of raw or half manufactured material into finished forms ready for final consumption.

The foregoing charts have been prepared on the basis of tables giving the actual facts in respect to the machinery, the product, and the wages of two successful cotton-mills, manufacturing what are known as standard sheetings, in New England. Technically these goods are known as 36-inch sheetings, No. 14's. In point of fact, the number of the yarn is a little coarser. The data have been combined so as to cover the entire period from 1830 to the present date, a part of them having been furnished from one mill and a part from the other. I have in my possession the accounts of many other cotton factories, and the statistics of the wages, covering a great variety of fabrics, during the last fifty years; but I have carefully chosen the data of two factories which have been uniformly successful, in which the capital stock has never been reduced, and of which the product has, to a large extent, been sold for export. This selection has been made in order that the data might not be affected in any measure beyond that of other occupations than cotton-spinning, by the many changes in the tariff which have been made since 1830.

In the main treatise of which this is an appendix, I have attempted to sustain the proposition that the rate of wages cannot be taken as a standard for determining the cost of production, even in money; but, on the contrary, that wages are a remainder over, or result of production, recovered from the sale of the goods, and subject to the prior claim for payment of the cost of materials and the profits of capital.

Wages will vary in rate in the same country, at different periods, in the same place; at the same period in different places; in different countries at the same time,—being determined by the distance of the factory from the source of the materials, by the intelligence and skill of the people who do the work, by the incidence of taxation, (the laws of different States varying on this point) and by many other elements which enter into the problem. On the other hand, although wages are deferred to profits, and are a remainder over, subject to deduction of profits from the sales, yet the competition of capital with capital not only always tends to a minimum of profit, but also to an increase of the product in ratio to the amount and effectiveness of the capital. Hence, while profits tend to a minimum, wages tend to a maximum. It therefore follows that, under these conditions, wages constitute an increasing proportion of an increasing product, provided markets can be found to take the increase without a reduction in price corresponding to the reduction in the labor which constitutes the true cost. In point of fact, very few nations have learned to apply machinery to the arts of life,—a larger portion of the population of the world is clad in homespun than in machine-made or factory-made fabrics. I have lately read a notice of a recent report, made in Manchester, to the effect that nearly 1,000,000,000 persons, out of a computed total of 1,400,000,000, may be considered as non-machine using nations, clad in hand-made fabrics, so far as they are clothed at all. In the United States, machinery is applied, on the whole, more effectively than anywhere else. Hence, although prices have diminished, they have not diminished as fast as the labor cost of production has been reduced. Consequently, wages have not only risen in rate, but also in purchasing power. All of this is proved by the figures of the charts which have been given above.

Between the two extreme dates which I have covered in the chart, 1830 and 1884, the cost in money for manufacturing a coarse cotton fabric has been reduced more than one half. In the same period, the rate of profit on each dollar invested, which sufficed to induce the construction of the factory, has also been reduced one half. In the same period, each unit of the machinery itself has become so much more effective, that one operative will perform three and a half times the work in eleven hours that one operative could perform, from 1830 to 1840, in thirteen hours. Thus it has happened that, while capital may now be satisfied with one quarter part as much money derived from the sale of the product as it formerly secured, wages have doubled per day, and more than doubled per hour, in the period named. From 1830 to 1840 inclusive, it was necessary to take fourteen per cent. from the gross sales of goods in order to pay ten per cent. on the capital of the factory. From 1880 to 1884 inclusive, six per cent. of the gross sales would suffice to pay ten per cent. upon the capital, while six per cent. profit would now be more nearly a normal rate.

In these charts I have treated the art of spinning and weaving cotton by machinery, upon what are called the self-acting mules, spinning-frames, and power looms. We may contrast the conditions of the same art, at the present time, in different parts of this country. In the heart of this country, upon the hill-sides and in the valleys of the great Allegheny region, in Virginia, in Kentucky, in Tennessee, and in the Carolinas, there is a population of two millions or more of people, who are even to this day chiefly clad in homespun fabrics, of which the yarn is spun upon the hand spinning-wheel, and woven upon the hand-loom. These people have been kept in isolation by the surrounding pall of slavery, until a very recent period. Their country is now being opened by railroads, and the art of making homespun fabrics will soon be a lost art among them. The capacity of five of these persons—to wit, two carders, two spinsters, and one weaver, in a day of eleven hours, is eight yards of coarse fabric, heavier, but of more open texture, and therefore more quickly woven by machinery than the standard sheeting. Five operatives in a modern factory would spin and weave one hundred-fold as much, or eight hundred yards a day. But we will limit the comparison to the actual product of standard sheetings, and we will assume that the home spinners could make eight yards of standard sheeting in a day. This would give them 2,400 yards as the product of a year, against 140,000 yards in the northern factory.

The cost of spinning and weaving the standard sheeting in the northern factory in 1883 was 1.08 cents per yard. If the southern operatives were obliged to sell their product in the open market at the same rate of wages—that is, at the wages which could be derived from 1.08 per yard, the total earnings of the five in one year would be $25.92, or a trifle over $5.00 each. If they were content with the profit on each yard which yields to the northern capitalist ten per cent. a year, it would be .43 of a cent a yard, or upon 2,400 yards $10.32. The total wages and profits of the five southern operatives, working by hand for one year, at the standard of cost and profit of the northern cotton-mill, would therefore amount to $36.24. On the other hand, in order that the earnings and profits of the southern operatives should be equal to those of the northern operatives and owners of the factories, it would be necessary that the homespun fabric should sell in the open market at about ninety cents a yard. It therefore follows that the high wages of the northern operatives are the result of the low cost of production, and that if the southern people now engaged in the art of homespun work can find other work to do, in dealing with the abundance of timber, in saving the wild fruits, in agriculture, or in the many other branches of work which their climate and soil open to them, but which are not open to the inhabitants of the Northern States, they will save both time and labor by an exchange of product, and by becoming inter-dependent, rather than by remaining isolated and independent. And this is what is now occurring. As soon as the incubus of slavery was removed and an exchange of products between the two sections of the country fairly began, each found that it could serve the other and and that slave-grown cotton was no longer king.

APPENDIX III.

In order to test the rule of the advance in the rates of wages which accompany improved methods of work and the substitution of machinery more or less automatic for hand work, I have compared the wages of two branches of industry employing men almost exclusively in special arts requiring a high degree of skill, to wit: the manufacture of pianos and the manufacture of edge tools.

In one piano factory of the highest reputation the rates of wages of five classes of workmen averaged

In 1843	$562 per year.
In 1880	824 " "

In another larger factory, the rates of wages of twelve classes of workmen have been as follows:

1853	$11.33 per week gold.
1860	12.23 " " "
1866	14.75 " " currency.
1872	18.00 " " "
1878	14.66 " " "
1880	17.50 " " gold.

In one establishment making table cutlery, eight classes of workmen averaged

1859	$1.50 per day.
1880	2.15 " "

In another on edge tools, ten classes of workmen averaged

1850	$1.60 per day.
1880	2.26 " "

In these examples the law of increasing wages is demonstrated, but there is no such unit in these arts as the standard sheeting, and I am unable to show how much the ratio of profit has diminished.

In fact, no other standard can be found like the standard sheeting, as it has been manufactured in precisely the same way since it was first introduced more than fifty years ago.

Even the statistics of the cost in money of the standard sheeting fail to show the true progress of the operatives. In 1830 and 1840 the machinery was much less automatic than it is now, and its operation called for a high grade of intelligence. From 1830 to 1850 the larger portion of the factory operatives were well-bred American women, graduates of the common schools, capable of writing the articles in the *Lowell Offering*. But to them the factory gave opportunity for progress, even though the hours of work were 13 to 14 per day and the work itself was arduous and continuous. The operatives who now earn nearly twice as much per day of 10 to 11 hours and more than twice as much per hour are, through no fault of their own, less instructed and less capable of doing work which requires versatility and individual capacity. They are mostly foreign-born. American women have gone up into more congenial employments at higher wages, which have been opened to them by the application of machinery to many arts which were mere handicrafts a few years since, and they have thus made room in the textile factories for the Canadian, Irish, English, and German immigrants, who now constitute the greater portion of the operatives.

Yet it will be observed that notwithstanding all these changes in the quality of the operatives, the improvement in the quality of the machinery has caused the share of the laborer to increase as steadily as the share of the capitalist has diminished; and this progress has continued in spite of all the chances and changes of meddlesome legislation.

APPENDIX IV.

Since this treaties was completed the invaluable report of the statistics of labor in Massachusetts for 1884, compiled by Carroll D. Wright, has been published.

It gives me another opportunity to prove the accuracy of my deductions.

In my treatise I worked from an *a priori* estimate of the value of the total product of the United States.

I deduced a value not exceeding ten thousand million dollars in the census year ending June 30, 1880, by estimating the several crops in quantity and in money. First.—By converting that portion of the wheat crop which is consumed in the United States into bread, and a large portion of the corn into meat ready for final consumption, and to this secondary or final form I applied the average retail prices. I also ascertained as nearly as possible the ultimate value of dairy products and the like. Second.—I converted the known quantity of textile fibres consumed within the limits of the United States, into fabrics, and I then estimated these fabrics at their value in finished clothing at the average prices which are charged by shopkeepers. Third.— I converted the known production of metals into machinery and other forms ready for final use, and valued them. Fourth.—I valued the timber product as furniture, dwelling-houses, and the like. Fifth.—I converted the sum of our imports into a value at its final point of consumption by estimating the cost of distribution and by other similar methods.

Of course this method is one which could not be made absolutely correct, especially by a private person working only in the intervals of active business. The conclusion was warranted in my own judgement by deductions from such facts as I could ascertain. I should not however have ventured to make use of this estimate in a scientific treatise, except its conclusions could be sustained by induction from the facts taken in detail.

By dividing my final estimate by the ascertained number of persons who were engaged, my *a priori* conclusion was that the average group of three persons, there being one person occupied for gain including the administrative as well as the executive force in each 2.90, would come into the possession of substance not exceeding in value $523, from which sum all profits, taxes, and wages must be derived.*

Upon a further analysis, a subdivision of this average sum which included ten per cent. estimated to be consumed directly upon the farms without going into the commercial stage, I found reason to assign to each one of those engaged in the work of administration, that is in the mental rather than the manual work, an income averaging between $1,000 and $1,100 a year, which being deducted left an average to each person engaged in the actual executive work of the country of between $430 and $450 a year.

It being assumed that each one of the latter class represented $2\tfrac{90}{100}$ persons, each person could enjoy only what $147 a year would buy, or in the last analysis what 40 cents a day will buy ; that is to say, if my estimate were correct, each member of a working man's family must find shelter, subsistence, clothing, and fuel on what 40 cents a day will buy, because such is the measure of the total product after setting aside five per cent. as the annual profit of the capitalist, and five per cent. more representing the small savings of the working people.

In other words my deductions *a priori* were, that the average share of the total product falling to each woman and child in the United States in the census year, including the domestic consumption of farmers' families, could not exceed what 55 cents a day would buy. Of this sum I assumed that 5 cents worth would be the domestic consumption of the agricultural population, leaving 50 cents a day as the average to each from that part of the production which was bought, sold, or exchanged. Five per cent. or 2½ cents a day set aside as the profits of capitalists, and five per cent. more or 2½ cents a day as the savings of the people, left 45 cents per day to be divided among the working people and the administrative force. Again subdividing this, and the apparent share falling to the family of each member of the administrative force seemed to be 90 cents to $1.00 per day, leaving to each member of each working man's family 40 cents.

All these computations were antecedent to any examination or test by consideration of actual rates of wages. They were deduced as the necessary result of the division of a total annual product estimated by entirely different methods than by computation of actual wages.

If, then, 40 cents a day be the average of the whole country, the proportion falling to the agricultural population, especially of the South and parts of the West, would be less. The proportion falling to the manufacturing population of the North would be greater. What then were the facts?

I have shown how far these figures coincided with the statistics of the average wages given in the United States census, and now have the satisfaction of comparing them with the facts elicited by Mr. Wright in the manufacturing State of Massachusetts.

Omitting common laborers, domestic servants, and the like, he has ascertained the average income of all persons engaged in various branches of manufacture to which machinery is applied in largest measure, or which require special skill. The list of occupations comprises the making of agricultural implements, of tools, boots and shoes, clothing, textile fabrics, furniture, persons engaged in the building trades, in the making of liquors, machinists, printers, makers of wooden-ware, and some other minor branches.

He finds the average wages of these persons in 1883, when they were somewhat higher than in the census year, to have been $10.31 per week, or $536.12 per year.

It will be observed that this list does not include the domestic servants, common laborers, or persons engaged in agriculture, even in Massachusetts itself, whose wages would bring down the average of the whole if they were included.

His results, even to this extent, may therefore be considered as fairly corresponding with the deductions made by myself, but the most conclusive proof of the accuracy of my deductions will be found in the treatment of what he calls the "budgets" of nineteen selected families assumed to represent the average of skilled workmen ; the expenditures of 400 families having been analyzed in the preceding year with which these "budgets" correspond.

These families comprise ninety-nine persons, of whom forty-one are engaged in some kind of gainful occupation—*i. e.*, earning wages. Each working member of this small force therefore represented a group of 2.17, as against the average of 2.90 in the whole country. The average income of each one of these persons was $372 a year, somewhat less than the average which I have assigned to each person in my estimate, but when we convert the $372 per year into so much a day for each person, the result gives forty-seven cents a day, in Massachusetts, in arts conducted mainly by machinery, against my estimate of forty cents for the average of the whole country, the group 2.17 being smaller.

It therefore follows that by every method of computation, and by every test which can be applied, my deductions are sustained.

It appears that even in the most prosperous State, the average income on which each person must subsist, find shelter, pay taxes, and make savings, even in arts requiring a high grade of skill, is less than fifty cents a day. If half the people of this country must live on what fifty cents a day will buy, the other half must live on what thirty cents a day will buy, since forty cents is the measure of all there is which can be assigned to their support ; yet, at this rate, Mr. Wright reports his conclusion that the standard of living of the workingmen of Massachusetts is in the ratio of 1.42 to 1 in Great Britain. He does not treat the condition of European continental laborers, but all students are well aware that the British workingman is better than his continental competitor in similar ratio.

The conclusion which can be drawn from these data, in my own judgment, would be this: Great Britain produces within its own limits but a moderate portion of the food of its people, and scarcely any of the materials used in its manufactures with the exception of iron, and is therefore forced to import by far the larger portion of its materials, and a large part of its food, and to pay the cost of freight thereon.

The various elements of her manufactures—in moderate part produced at home, and in large part brought from other countries—are then combined into an annual product of a certain value, out of which rents, profits, taxes, and wages must be derived. Under present conditions the remainder over, left to the British workingmen, as compared to the Massachusetts workingmen, is in the ratio of about two to three, *i. e.*, the Massachusetts working man or woman is fifty per cent. better off than the British working man or woman.

Upon the continent, where the resources of the several countries themselves are even less in ratio to the number of persons to be sustained, the value of the annual product is less in proportion than it is in Great Britain; while the labor exerted is very much greater than it is either in Great Britain or in in this country. Consequently the remainder over, after paying for the enormous cost of standing armies, and after being subjected to the withdrawal of one man in twenty from the productive work, is less probably by one half than it is in this country, and by one third than it is in Great Britain.

As a natural consequence large masses of people in Italy, in Germany, and in some parts of France and Belgium, barely exist upon the edge of starvation.

It seems almost a necessity to bring this matter down to the unit of the individual, in order that the people dwelling upon the continent of North America may in some measure comprehend the advantage of the position, and of their freedom from vested wrongs under which their fellows are suffering in countries of so-called older civilization.

Another branch of the subject to which I have as yet given little attention, needs to be explored, in order to show that 40 cents worth is enough for moderate comfort, if it is used with moderate intelligence; for instance, the jail of the county in which I live is admirably conducted. The prisoners are adults, boys and girls being sent to reformatories. The food of these prisoners consists of bread made from the best flour, and the meat consists of the remainder of the carcass of the best beeves and other animals after the fine cuts have been taken off for the first class hotels. These persons are served with a moderate quantity of tea, with rye coffee, with such vegetables as are suitable; in short, with an abundance of food, and it is probably better cooked than in the average family of common laborers, and yet the prime cost of the provisions required by each prisoner, delivered at the jail, is but a trifle over 12 cents per day; of course it is prepared by prisoners.

Now it appears, both from Mr. Wright's investigations and from those of Dr. Engel, of Berlin, that the cost of subsistence of a workman's family, earning from $300 to $750 a year, is sixty per cent. of his whole expenditure.

If, then, an abundant supply of nutritious food for an adult can be procured in Massachusetts at a cost of $50 a year, and the same economy could be used in respect to other items of expense, an income of $90 a year to each person would suffice for wholesome conditions, while $100 a year would amply provide for the excess of rent which working people in Massachusetts are obliged to pay above their competitors in England.

This latter assignment of $100 a year to each person,—which was the average of 20 or 30 years ago,—would be a fraction under 28 cents a day. There can be little doubt that the rate of wages has advanced in this ratio, *i. e.*, from 28 to 40 cents per day for each person during the last twenty years, and that each dollar has also greater purchasing power.

If, then, the margin be narrow and if want treads still close upon the steps of welfare—courage may yet be taken as to the future under the application of the law of diminishing profits and increasing wages.

In conclusion, at the risk of repetition, let me again call attention to the fact, that in order that each person of the present population of the United States, computed at this date at fifty-eight million, may enjoy five cents worth per day more than the average assumed in this treatise, it would be necessary that the production of each person should be increased $18.25 per year; or, in other words, that each person in a group of three engaged in gainful occupation should produce $55 worth more than each one now produces, and find a market for the increasing product without diminishing prices.

Now, $18.25 per person, multiplied by fifty-eight million, gives an aggregate of $1,058,500,000. This sum is twice and a half the value of the present wheat crop of the United States, ten times the value of the pig-iron produced in the United States, about double the value of all the textile fabrics; or, to put it in another way, the people who are now at work, numbering at the proportion which the working force of the census year bore to the whole, about twenty million, must add to the present product the value of our wheat crop, say $350,000,000; to the value of our pig-iron product, say $90,000,000; to the value of all our production of textile fabrics, say $650,000,000, total, $1,050,000,000, and must find a market for the sale of the increased product, in order that each one of their number may earn fifteen cents a day more than they now do, and that each one of a group of three may be able to consume more than they do now by what five cents a day more will buy.

In this view of the matter, progress in material welfare is and must be very slow.

This problem is commended to all who expect to improve the welfare of the people by changes in respect to land tenure, or by creating paper money, or "fiat money," or by compulsorily shortening the hours of labor, and by other methods of meddlesome interference, by statute, with customs which have been gradually evolved during the last two centuries.

May we not respectfully suggest that such progress can be accomplished only by the advancement of science, beginning in the common schools, with manual and technical instruction as well as with mental work.

Increased production and a wider market constitute the only sources from which the money can be obtained by which the rate of wages can be advanced.

On the other hand, there must be this increase of production, in order that even if the rate of wages is not advanced, each unit of the wages will buy as much as it now does.

The true function of commerce must be fully comprehended in order that such an advance may be speedily reached. It cannot be reached until the present fallacies in regard to wages have been given up, nor until the principle shall be accepted that high rates of wages, expressed in terms of money, are the result of low cost of labor, expressed in hours or efforts.

In the great competition under which service for service is rendered, those nations which apply machinery to the fullest extent, and to the most adequate resources, make the largest product at the least cost of labor.

In their exchanges with what are called the barbarous or hand-working states of the world, or with those nations in which machinery has been applied to the arts in least measure, they gain the most for themselves, while rendering the greatest service to those with whom they deal.

This is the secret of English wealth.

This is the secret of the higher wages of the English-speaking people.

This is the secret which the people of the United States have yet but half comprehended, because the abundance of their product is so great, that no stress of want has yet compelled attention to be given to the science of political economy, and to the methods by which the burdens of taxation can be most easily borne.

This subject is a vast one; it includes not only the tariff question, but also the much more complex and difficult question of local or municipal taxation, in respect to which there is no uniform system or practice in the United States.

If I have succeeded in calling attention to the fundamental principles which must be considered before we can even begin to deal intelligently with these vast social questions, I shall have accomplished my purpose.

It will have been apparent to the reader that in this treatise and its appendices, two separate lines of investigation have been followed.

In the first place the principle has been laid down that by way of, or by force of, competition, there is a tendency in the rate of profit, interest, rent, or by whatever name or designation the share of the capitalist is defined, to diminish.

On the other hand, there is a tendency in the rate and purchasing power of wages to increase.

These tendencies are subject to variation in short periods of time, owing to short crops, war, or other similar causes, but in any long period of time they become rules.

Furthermore, in any country inhabited by a substantially homogeneous people, high wages both in rate and in purchasing power are the necessary consequence or result of the low labor cost of production.

This rule will also apply between different countries subject to variation arising in the practice of hereditary arts, or from the imposition of customs duties and other like causes.

This rule is also subject to temporary variation—but in a long period of time may be considered absolute in its working. These positions have been sustained historically and by the citation of facts growing out of existing conditions in the United States, and they form the main purpose of the treatise.

The second subject—or division of the main subject as it might perhaps better be called—consists in the attempt to measure the annual product of the United States in terms of money, and thereby to determine the possible share or remainder enuring to those who do the work, by which measure the average rate of wages in the United States at a given time or at the present time may be established.

This part of the subject is sustained by such testimony as may be available from official documents, but must be considered as only approximate in its terms ; suggesting a method whereby these facts may be hereafter determined rather than a conclusive treatise upon present conditions which it would be impossible for a private person to work out in an absolutely certain manner.

It may be readily conceived that the Government Bureau of Statistics, or the officers of the next census could make a very accurate computation of value of our annual product by first ascertaining the value of grain, cotton, metals, timber, wool, and other fibres and the like, and then tracing each subject through its various conversions to the point of ultimate consumption, as bread, clothing, shelter, machinery, etc., etc.—the value of that portion exported being very easily computed separately.

Of course there would be some errors and omissions, but they would balance each other, and the result in cents per day per person, or dollars per year per family, would be but little affected by the sum of all probable errors.

APPENDIX V.

My attention having been called to a computation of the value of the annual product of the census year, which is included in the report of Mr. Joseph Nimmo, Jr., Chief of the Bureau of Statistics, for 1884, I have requested him to give me the data upon which he reached his conclusions in the matter, and I have the satisfaction of submitting his letter herewith.

BUREAU OF STATISTICS,
Washington, D. C., October 21, 1884.

DEAR SIR :—Your letter of the 17th inst. has been received, and I will reply as follows in regard to the total value of our annual product :

The estimate of $3,600,000,000 for the product of agriculture was given to me by Mr. J. R. Dodge, a year ago, as the result of a series of careful investigations, and he firmly adheres to that estimate. Mr. Dodge had charge of the census agricultural statistics, and I regard him as the best authority in the United States upon that subject.

The following is a foot-note upon this subject, which appears in my article on "American Manufactures," contributed to the *North American Review*, of June, 1883, and is taken from a memorandum upon the subject given to me by Mr. Dodge :

" This is an estimate made by Mr. J. R. Dodge, Statistician of the Department of Agriculture, and Special Agent of the Census for the Collection of Statistics in regard to Agriculture. The census gives $2,213,402,564 as the estimated value of farm productions. This, however, does not include the increased value of live stock, nor the value of the products of pasturage on the public lands. It also omits to a very large extent products of horticulture."

All the other values, in making up the aggregate, are directly from the Census Office ; so that my total of $9,817,900,652 in the foot-note on page 40 of my annual report was made up as follows :

Agriculture	$3,600,000,000
Manufactures	5,369,579,191
Illuminating gas (partly estimated)	30,000,000
Mining	236,275,408
Forestry (partly estimated)	455,000,000
Fisheries	43,046,053
Meat production and wool clip of ranches (estimated)	40,000,000
Petroleum—manufactured product	44,000,000
Total (materials out)	$9,817,900,652

I conferred very fully with the Acting Superintendent of the Census, Mr. Geo. W. Richards, an exceedingly intelligent and able man, who appears to have a thorough understanding of the whole census figures. Regarding the total value of the products of manufacture, he stated to me that while there are some duplications in it, the omissions amount to very much more. It is certain that the values are, on the average, below the actual values ; and that there is a considerable amount overlooked ; besides, the census did not take into account the products of any establishment the value of which products was less than $500.00.

I have no doubt that the total value of the products of all industries was over rather than under $10,000,000,000, perhaps in very considerable measure, but of course there are no exact data beyond those given in the census. We may safely say on the basis of the census data that the total value of the products of all industries in the United States was at least ten thousand million dollars.

I am, sir,
Very respectfully yours,
Jos. Nimmo, Jr.,
Chief of Bureau.

This computation, it will be seen, is almost identical with my own, except that Mr. Nimmo uses the expression "at least," where I have said that the annual product in the census year was " at most," $10,000,000,000.[1]

---

[1] It is, of course, impossible to bring such a problem as this to very exact terms by an unofficial investigation ; but if, however, we assume an error of five per cent. in the computation of the gross value of the annual product, such an addition would be substantially two and a half cents a day to each person, and would amount to the gross sum of $500,000,000 a year on the average population of the last four years.

Such an addition would fully cover the point in respect to which there are n*f* actual data in the census or elsewhere, but which must be treated wholly as *a* matter of observation and judgment, to wit : the steadily increasing proportion of prosperous persons who may be economically called the well-to-do, or in common speech the forehanded men ; such as prosperous shopkeepers, able foremen in the mechanic arts, farmers whose principal tools are their own brains, capable women taking part in occupations formerly controlled wholly by men, small manufacturers who own and control their own works,—and the like. In the sorting which I have previously made on a broad and general scale, I have, perhaps, left no place for this class of persons, but by adding five per cent. to the assumed product of $10,000,000,000 in the census year full provision would be made for them in the following classification :

Total production as first computed	$10,000,000,000
Domestic consumption on farms and domestic product of families which is not exchanged or does not come into the commercial product	1,000,000,000
Commercial product	$9,000,000,000
Share of capitalists, 5 per cent.	$450,000,000
Savings of the people, 5 per cent.	450,000,000
Addition to the capital or wealth of the nation	900,000,000
Wages Fund	$8,100,000,000

Share of 1,100,000 persons who are assumed to be engaged in mental and administrative work, computed at $1,000 each, including 227,210 teachers and scientific persons. This class may be subdivided as follows :
200,000 teachers in the lower-grade schools, scientists, authors, artists, young lawyers and clergymen, or other persons

Attention may be called to the proof which is to be found in Mr. Nimmo's excellent annual report of the actual and necessary preponderance of domestic as compared to foreign commerce.

It will be very apparent to any one who considers the statistics of these classes at $550—$110,000,000. 900,000 merchants, tradesmen, officials or others in the higher work of administration at $1,100 each—$990,000,000 . . $1,100,000,000
16,200,000 farmers, laborers, mechanics, artizans, operators, clerks, dress-makers, and other wage-earners, $432 cach   7,000,000,000

$3,100,000,000

Total assumed product thus accounted for as above . $10,000,000,000
Add 5 per cent upon this gross product in order to account for the larger consumption of well-to-do farmers, foremen, prosperous country tradesmen or shopkeepers, and other classes, of whom there may be one million, and to each of whom $500 each above the average might be assigned. Such an assignment would give five per cent. of the farmers, or 200,000, a cash income of $932 each, in place of an average of $432, and would bring 800,000 of those who have been classed as tradesmen, mechanics, operatives, clerks, etc., from $432 up to $932 each . $500,000,000

Total $10,500,000,000

If any larger product should be assumed it would be difficult to trace it either in the form of greater savings or in larger consumption. No evidence can be found of any larger addition to capital than has been given, and no trace of higher wages so far as the census returns cover rates of wages; but the incomes of what may be called the prosperous middle class, to whose consumption the possible additional product has been assigned, are not to be found in any statistical returns.

If such an addition ought to be made, then the average product of each person in the census year was 57½ cents per day, and the addition of 2½ cents to each person per day is to be added to the previous computation of 55 cents.

This reasoning is based upon the position taken in this whole treatise, to wit: that the progress of the few is not at the cost of the poverty of the many; but, on the contrary, the ever increasing abundance which has been produced or brought forth to the use of men in recent years, may be shared by all classes according to the relative capacity, integrity, and industry of the respective

of domestic agriculture and manufactures, that only a very small portion of the products of agriculture could be imported from any other country—mainly consisting of sugar, rice, a portion of our necessary supply of wool, and a very few other articles,—while of necessity a very considerable portion of agricultural products are raised within each State itself. It is also true that by far the greatest proportion of the mechanical and manufacturing arts exist from necessity and not from choice, within the limits of particular sections of the country and even in particular States.

Reference has been made in the body of the treatise to the way in which special arts have become rooted or centralized in particular places, sometimes without any apparent reason, except that groups of population have become habituated to the practice of such arts, so that they have become hereditary. Under such conditions the law of decreasing relative profits and increasing relative wages can be observed in the clearest manner, as well as the rule of high rates of wages accompanying or resulting from low labor cost of production, because in such places all the subsidiary employments have gathered around the chief centre, and every possible facility exists for making the largest product by means of the work of the least number of persons.

The interdependence of agriculture and of the manufacturing and mechanical arts, and the necessary proportion of each in every prosperous State, are proved in a very skilful manner, by Mr. J. R. Dodge, the able Statistician of the Department of Agriculture of the United States, by means of a series of diagrams contained in the Agricultural Report for 1883, showing the manner in which the values of farms and farm products are influenced by the establishment of various branches of the mechanic arts and of the lesser manufactures in their immediate proximity.

---

members of each class; *provided* the functions of the legislators are limited to such acts as may leave the principle of competition in the use of land and of all its products as free to work out its just results as the protection of the young, the ignorant, or the incapable from injustice will permit.

In other words *competition* leads of necessity to the most effective and beneficent system of *co-operation* among men.

COMPARISON OF THE QUANTITIES AND FARM VALUES OF THE PRODUCTS OF AGRICULTURE OF THE UNITED STATES, IN 1859 AND 1879.

Product.		1859.			1879.		
		Quantity.	Price.	Value.	Quantity.	Price.	Value.
Meats				$300,000,000			$800,000,000
Poultry and Eggs				75,000,000			180,000,000
Butter	pounds	500,000,000	$0.16	80,000,000	900,000,000	$0.21	189,000,000
Cheese	"	130,000,000	.095	12,350,000	300,000,000	.095	28,500,000
Milk consumed	gallons	1,000,000,000	.06	60,000,000	1,800,000,000	.075	135,000,000
				527,350,000			1,332,500,000
Corn	bushels	837,792,740	.43	360,680,878	1,754,591,676	.396	694,818,304
Wheat	"	173,104,924	.72	124,635,545	459,483,137	.951	436,968,463
Oats	"	172,643,185	.25	43,160,796	407,858,999	.36	146,829,240
Rye	"	21,101,380	.52	10,972,718	19,837,595	.756	14,992,686
Barley	"	15,825,898	.55	8,704,244	43,997,495	.666	29,302,332
Buckwheat	"	17,571,818	.58	10,191,654	11,817,327	.594	7,019,492
Rice	pounds	187,167,032	.035	6,550,846	110,131,373	.06	6,607,882
				564,896,681			1,336,538,399
Hay	tons	19,083,896	8.00	152,671,168	35,150,711	11.65	409,505,783
Irish Potatoes	bushels	111,148,867	.40	44,459,547	169,458,539	.483	81,848,474
Sweet Potatoes	"	42,095,026	.40	16,838,010	33,378,693	.45	15,020,412
Peas and Beans	"	15,061,995	1.33	20,032,453	9,590,027	1.50	14,385,041
Market Garden Products				16,159,498			21,761,250
Orchard Products				19,991,885			50,876,154
Hops	pounds	10,991,996	.07	769,440	26,546,378	.24	6,371,131
				118,250,833			190,262,462

THE RATE OF WAGES? 145

Product.		1859.			1879.		
		Quantity.	Price.	Value.	Quantity.	Price.	Value.
Cane Sugar	. hogsheads	230,982	$85.00	$19,633,470	178,872	$90.00	$16,098,480
Maple Sugar	. pounds	40,120,205	.12	4,814,425	36,576,061	.13	4,754,888
Cane Molasses	. gallons	14,963,996	.30	4,489,199	16,573,273	.35	5,800,646
Sorghum Syrup	"	6,749,123	.30	2,024,737	28,444,202	.33	9,386,587
Maple Syrup	"	1,597,589	.80	1,278,071	11,796,048	1.00	1,796,048
Honey	. pounds	25,366,357	.20	4,673,271	25,743,208	.22	5,663,506
				36,913,173			43,500,155
Cotton	. pounds	2,274,372,309	.093	211,516,625	2,771,797,156	.098	271,636,121
Wool	"	75,000,000	.26	19,500,000	240,681,751	.28	67,390,890
Hemp	. tons	74,493	190.00	14,153,670	5,025	200.00	1,005,000
Flax	. pounds	4,720,145	.20	944,029	1,555,546	.25	391,387
				246,114,324			340,423,398
Tobacco	. pounds	434,209,461	.05	21,710,473	472,661,157	.085	38,758,215
Flax Seed	. bushels	566,867	1.00	566,867	7,170,951	1.25	8,963,689
Grass Seed	"	900,040	1.40	1,260,056	1,317,701	1.50	1,976,552
Clover Seed	"	956,188	5.00	4,780,940	1,922,982	6.00	11,537,892
				6,607,863			22,478,133
Wines	. gallons	1,627,242	.50	813,621	20,000,000	.60	12,000,000
Beeswax	. pounds	1,322,787	.30	396,836	1,105,689	.33	364,877
Aggregate values . .				1,675,725,972			3,726,331,422

Export in 1880 mainly from crops of 1879. Products of agriculture, $685,961,091.

By the courtesy of Mr. Dodge, I am enabled to give his corrected estimate of the value of farm products for the year covered by the census returns. It will be observed that his estimate somewhat exceeds my own, even though he does not include the domestic farm consumption of fruit and vegetables, but inasmuch as a very considerable part of the hay and corn crops are converted into meat and dairy products, a fair allowance for this duplication would bring the two estimates almost to an exact agreement.

The great increase both in quantities and in values between the years 1859 and 1879 will be observed, but although the farm value was greater for the same quantities in 1879 than in 1859 it by no means follows that consumers at a distance paid more for grain or dairy products; the advance in prices at the places of production, so far as it can be traced, is less than the reduction which was made between those two dates in the charges for moving those products by railway from producer to consumer. The extension of the railways to new lands first made production and sale possible; then as production increased the reduction in the railway charge occurred, so that it has not been until the present year, 1884, that any material reduction of price has been felt by farmers, and even in this year this reduction has only occurred in any great measure with respect to wheat and wool. (See next essay on the Railway, the Farmer, and the Public.)

But this increase of the products of agriculture has been accompanied by an extension of manufacturing and the mechanic arts.

It is, of course, impossible that any community can long exist which is exclusively devoted to agriculture, except under a system of slavery. The artisan must accompany the farmer almost from the beginning of the settlement of any section; next, or almost at the same time, comes the minister, lawyer, doctor, shopkeeper, domestic servant, and laborer; soon after, or in the later years, even before the farmer, comes the railway with its employés, and presently the factory of some kind, each following a natural order and sequence, except when interfered with by restrictive statutes limiting the freedom of labor as in the days of slavery, or indirectly preventing commerce between the States.

The most perfect example of the working of this law is to be found in the rapid growth of various branches of manufacturing industry in the Southern States since the statutes imposing slavery upon that section were removed. Here was a section almost wholly agricultural : its people were dependent upon the North even for pots and pans ; for clumsy "nigger" hoes and other rude and heavy implements of agriculture, fit only for slaves to use ; for wagons ; for all their iron ; and also even for hay, corn, and bacon. Yet the moment the burthen of slavery was removed, all the arts sprang into existence,—some of them, perhaps, prematurely. In many branches of industry the tide of commerce is reversed : the largest single tannery in the country gets itself established in Tennessee, and sends its leather to New York ; Alabama discovers the imperial deposit of iron and coal of the world among her pine woods, and sends her product to New England ; the mountain section sends its hard woods in various half-manufactured forms all over the North and West ; and in every direction the interdependence of agriculture, manufacturing, and commerce, asserts itself as the natural outcome of liberty.

But in the so-called farming States of the West, the necessary and almost simultaneous growth of all the arts of life is most apparent. As an example of the evolution of industrial society, no better example can be taken than the State of Ohio, lying midway between East and West. Within a generation Ohio was rated as almost exclusively devoted to agriculture. Even as late as 1869 nearly one half of the small traffic on her railways was merely *through* traffic, in which the State itself had little interest. In 1883 a vast change had occurred which may be pictured as follows :

Ohio lies midway between East and West. In 1883 it contained 6,897 miles of railroad, against 3,324 in 1869. In 1869, the actual tons moved over all the railways reporting in the State numbered 14,559,704, of which fifty-five per cent. represented local traffic and forty-five per cent. through traffic. In 1883, 63,683,643 tons were moved, of which 66⅔ per cent. represented local traffic and only 33⅓ per cent. through traffic, showing how the local traffic gains, both absolutely and relatively. The charge per ton per mile in 1869 was 2.446 cents; in 1883, only .875 cents per ton per mile. Graphically the Ohio Railroad traffic may be represented in this way:

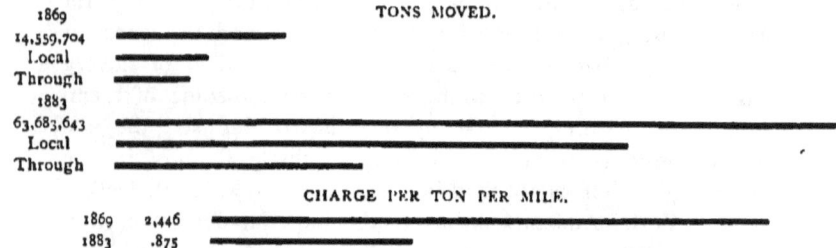

"The actual freight charge on all the railroads reporting in Ohio in 1883 was, in round figures, $67,000,000. Had this traffic been subjected to the charge of 1869 the sum would have been $201,800,000.

"The difference between these two sums is, in currency, $134,-800,000; in gold, $89,400,000. Now since two thirds of this traffic was *local* traffic, the saving in rates to the people of Ohio since 1869, on their local traffic only, was, in currency, $90,000,-000; in gold, $60,000,000."—From "The Railway, the Farmer, and the Public," reprinted herewith.

The saving which ensued in a single year growing out of the application of capital to railways, therefore, either added sixty million dollars to profits and wages or else it saved as much labor as would be represented by that sum in the work of subsisting, clothing, and sheltering the people of the State.

Now what have been the forces that have worked this great change? What caused the railways to be built, and what new conditions have the railways brought into existence? How do these new conditions themselves react in sustaining the railways by giving them this extraordinary increase in local traffic?

In order to understand this matter fully it would only be necessary for an acute observer to compare the relative conditions of the people of Ohio with an equal number who now exist in Eastern Kentucky and Tennessee and in Western North and South Carolina under conditions similar to those of a century ago in other parts of the country.

But in the absence of such actual observations we must again resort to statistics which prove the beneficent law of interdependence as compared to the independence and isolation of the mountaineers.

For this purpose the four principal subdivisions of the census should be increased to seven.

Table of all persons occupied in gainful occupations by the census of 1880:

Class 1.—Persons engaged in agriculture, including farm laborers	7,670,493
Class 2.—Professional and personal service, omitting laborers not specified	2,215,015
Class 3.—Trade and transportation	1,810,256
[1] Class 4.—Pursuits which are mechanical rather than manufacturing, according to common custom in classifying them	2,397,112
Class 5.—Pursuits which are of the nature of manufacturing rather than mechanical, according to common custom in classifying them, by estimate	1,200,000
Class 6.—Mining and pursuits immediately connected therewith, separated by estimate	240,000
	15,532,876
Class 7.—Laborers not specified, who are doubtless distributed in the service of the various arts or occupations included in the last five classes—agricultural laborers having been separately enumerated—but doubtless many laborers pass from one to another class as occasion may require	1,859,223
	17,392,099

[1] Judgments will vary in making this subdivision. I have classified machinists, for instance, numbering 101,130, as being in the factory division, and I have placed milliners, dress-makers, and sempstresses, 285,401, as well as tailors and taileresses, 133,756, on the mechanical side, although much of the

Perhaps we may account more fully for the progress of Ohio by considering the ratio which each class of occupations of the people now bears to the other in that State.

For this purpose we may sort them according to the census of 1880. The population in that year numbered 3,198,062, of whom 994,475 were engaged in some kind of gainful occupation, comprising 1 in 3.22 as follows:

Agriculture	399,495
Professional and personal service	250,371
Trade and transportation	104,315
Manufacturing, mechanical, and mining	242,294
	996,475

The principal subdivisons of the latter class will be found in the following lists, and it will be observed that by far the larger part of these arts exist in Ohio in the nature of things; they have grown out of the necessary diversity of occupations which has ensued from the application of science and invention to all the arts of life.

Tailors, dress-makers, and seamstresses	33,212
Carpenters and joiners	29,770
Blacksmiths	14,623

work of making clothing is now done in workshops which might well be designated as factories. These latter classes differ however from textile factories in this respect: that workshops for the manufacture of clothing by women are apt to be established at centres where large numbers of men are congregated who are engaged in other work, as in Chicago and other Western cities in recent years.

If all those whose occupations tend to concentration in factories were classed as manufacturing operatives, including clothing factories, hat factories, metal-goods factories, textile factories, and the like, the proportion classed as manufacturing would probably be about even as compared to those engaged in the mechanic arts—*i. e.*, in round figures:

Manufacturing	1,800,000
Mechanical	1,800,000
Laborers taken over from personal service as auxiliaries in these arts, say	400,000
Total	4,000,000

Iron- and steel-workers	13,419
Painters and varnishers	11,458
Boot- and shoe-makers	10,964
Brick and stone masons and stone-cutters	10,713
Machinists	7,498
Carriage, car, and wagon makers	7,020
Engineers and firemen	5,860
Butchers	5,713
Cabinet-makers and upholsterers	5,615
Miners	5,575
Coopers	5,357
Cigar makers and tobacco workers	5,297
Printers	4,658
Saw-mill operatives	4,148
Millers	3,919
Manufacturers and Officials' Manufacturing Cos	3,811
Harness, saddles, and trunks	3,661
Apprentices	3,525
Brick and tile makers	3,355
Tinners	3,331
Bakers	2,983
Cotton, wool, and silk	1,818
Brewers and malsters	1,744
Gold, silver, and jewelry	1,260
Wheelwrights	1,028
	211,335
Unenumerated, or less than 1,000 each	30,959
	242,294

It needs but a glance over the titles of these manufacturing and mechanical occupations to see that, given a considerable area of fertile land and an intelligent and free system of agriculture, nearly all the other occupations in this list must of necessity follow or accompany agricultural development; while most of these occupations, especially those of mechanical industry, must not only exist within the State itself, but must concentrate in and around every populous centre of the State, because the work is of such a kind that it cannot be imported from any other place except at a greater cost.

Towns and cities grow—they are not made,—and few men can even foresee by a few years where they must exist; but where they have grown they serve the agricultural population around them and are served by them. Out of this exchange comes increased welfare, and both city lots and country farms increase in value as the result of the facility which is given by their proximity for attaining the best conditions of life with the least effort, *i. e.*, a less quantity of labor and a greater quantity of products resulting in lower cost of production and higher rates of wages. The diagrams given by Mr. Dodge furnish a very interesting proof of this necessary co-existence in every State, of agriculture and the special mechanical and manufacturing arts which give employment to the largest number of persons and which must accompany agriculture.

It will be observed that in Ohio the proportionate occupation of the people is as follows:

Agriculture	41 per cent.
Professional and personal service	25 "
Trade and transportation	10 "
Manufacturing, mechanical, and mining	24 "
	100

If we apply this analysis to one of the youngest of our States, which is assumed to be devoted almost exclusively to agriculture —Oregon—we again find an example of diversity of occupation which proves how necessary all the arts are to any State, even if there are no great factories within its limits.

The population of Oregon in 1880 was 174,768, of whom 67,343 were occupied in gainful work in the following proportions, or 1 in 2.60:

Agriculture	27,091	40.3 per cent.
Professional and personal service	16,645	24.7 "
Trade and transportation	6,149	9 "
Manufacturing, mechanical, and mining	17,458	26 "
	67,343	100

Another example may be found in Kansas, another young State, as yet devoted mainly to agriculture:

Population in 1880, 996,096. Occupied, 528,302, or 1 in 1.90 (witness the very high ratio of workers).

Agriculture	303,557	57.5 per cent.
Professional and personal service	50,872	9.5 "
Trade and transportation	103,932	20 "
Manufacturing, mechanical, and mining	69,941	13 "
	528,302	100

In this State the railroad opened the way, or preceded agriculture, and the true balance of occupations has not yet become adjusted, but when families increase and the true balance of population is attained the same proportions will doubtless be reached as in Illinois and Indiana or other prairie States.

In the whole United States the proportions were as follows :

Agriculture	7,670,493	44 per cent.
Professional and personal	4,074,238	23.5 ,,
Trade and transportation	1,810,256	10.5 ,,
Manufacturing, mechanical, and mining	3,837,112	22 ,,
	17,392,099	100

In the great States in which diversified industry has been developed most freely and fully, like Ohio, Indiana, Illinois, and Wisconsin, the proportions of the occupations of the people substantially agree with the average of the whole country ; while in the South, where all diversity was forbidden by slavery, a rude kind of agriculture was until lately the rule, and the true diversity of free civilization is just beginning to assert itself. On the other hand, on the sterile soil of the Eastern States, in a climate in which indoor or factory occupations are most consistent with comfort and welfare, the manufacturing and the mechanic arts assume the preponderance that agriculture possesses elsewhere, while what little good arable land there is possesses the highest value.

By these subdivisions of labor the quantity of labor is diminished, and the quantity of product is increased ; then, as transportation becomes less and less costly exchanges cover a wider area. Each State, and each section of a State, therefore, takes up the work for which its soil and its people are best adapted, and in that State or section in which the best conditions are to be found, the sum recovered from the sale of its products will yield the largest profit and the highest wages, corresponding to the low cost in the labor in the work done.

If each State could be content to work out its just results in this way, there would be less contention ; but, unfortunately, the representatives of a few very much concentrated interests arrogate to themselves an importance which becomes somewhat ludicrous when subjected to comparison with others that excite little attention.

For instance, the whole country is now disturbed : commerce, both national and international, is adversely affected ; constructive enterprise is checked ; large numbers of people are thrown out of employment, while wages are consequently depressed,— simply by the continued coinage of light-weight silver dollars under the present act of coinage.

The purchase of silver bullion for this coinage is continued at the instance of what are known as the silver-producing States, in order to sustain the so-called "silver interests" of the country.

The value of the silver produced during the last few years, measured by comparison with the standard of gold has been about forty million dollars a year.

Under an act of Congress, more than half this product of silver is purchased by the Treasury in the form of silver bullion, and is coined into light-weight dollars, which are not wanted for use, and which are stored in costly vaults. The average tax which is imposed upon each person for this purpose is a little over forty cents a year ; each voter's proportion is about two dollars and a half a year. Perhaps the voters of this country are too busy to pay much attention to so small a perversion of the powers of Congress, or to remedy a wrong that only costs twenty-four million dollars a year, and which is imposed upon them in order to support a private interest. It may, however, be well to assign a ratable proportion of this tax to some of the towns and cities of the country, in order to show their share of this burthen :

New York City pays about	$560,000
Philadelphia " "	400,000
Chicago " "	240,000
Boston " "	160,000
My own little town of Brookline, Mass., pays about	4,000

This purchase of silver at the cost of the taxpayers, stimulates a product which is not wanted and which it would be desirable to leave to the working of the natural laws of trade, in order that the true ratio of silver to gold, *i. e.*, the true value of silver in terms of gold, may be determined. This cannot happen so long as the United States Government "bulls the market," if one may use the slang of the street. This measure is as obnoxious to the bi-metallist as it is to the advocate of the single standard of gold.

If the dangerous nature of our present course cannot be forced upon public attention by argument, it may be well to try another method. Let us measure the importance of the silver interest, so-called, by a comparison with some of the other products of our mines and of our agriculture, and for this purpose we will first compare the relative importance of the *silver mines* and of the *hen yards* of the country.

The census valuation of eggs and poultry was far below that of the experts who compile the annual data of our poultry and dairy products, but assuming that our hen population has increased in the same ratio as our human population, our annual supply of eggs is over five hundred million dozen, which at the low price of sixteen cents a dozen would be worth $80,000,000.

The product of what we may call our "*hen industry*" is therefore twice that of our *silver mines*, and it is immeasurably more important, because the proceeds of the sales are enjoyed by the least wealthy portion of our farming population, while the proceeds of the silver mines have in great measure gone to build up a few "bonanza fortunes," or have been wasted in vain attempts to increase an excessive and comparatively useless product of the same metal.

The most competent judge in this country of the cost of silver, the owner of the largest silver-ore reduction works in the world, (who never owned but one silver mine, in which he lost every cent which he put into it,) lately gave me his deliberate opinion that every dollar's worth of our present silver product cost the country not less than two dollars in gold.

But let us compare with another metal. Iron lies at the foundation of all the arts—it is immensely more important than silver The producers of iron are struggling under adverse conditions with no such purchaser as the United States, of two million dollars' worth a month to sustain their market; but the value of the product of our iron mines is more than double that of silver, and may be computed in this year of depression at not less than $90,000,000. Why not buy $2,000,000 worth of pig-iron per month and store it in some other vaults?

Wool, again, is one of our lesser farm products; it, like silver, has been stimulated by legislation to the point of an apparent excess of production of those varieties which can be raised in this country, so that the price is very low, but the clip of this year, which now comes to market mostly in an unwashed condition, is yet worth fifty per cent. more than the silver product, the clip of 1884 being computed at 320,000,000 lbs., which at 20 cts. is worth $64,000,000. Why not buy $2,000,000 worth of wool per month at the cost of the tax payers and thus stop the slaughter of sheep?

The estimates of our dairy products adopted by Mr. J. R. Dodge of the Department of Agriculture give the value of milk, butter, and cheese at $350,000,000, or about nine times the value of silver.

But perhaps the impudence of the demand of the silver interest can be pictured best by a graphical illustration, which will bring the relative importance of the respective products which I have cited into clearest view. I will give my own computation of the value of the products of the hen yards in 1884 based on the census of 1880, and also the commercial valuation of poultry and eggs adopted by Mr. Dodge, which are now computed at over $180,000,000 per year.

The parallelogram on the next page, enclosing separate graphical comparisons of these several products, represents the value of the annual product of 1884 on the basis of the previous computations for 1880, estimated at $11,400,000,000. The respective values of silver, pig-iron, wool, and dairy products are drawn on the same scale as the outer parallelogram.

Total product of the United States in 1884 computed at $11,400,000,000 — indicated by the outer line of the main parallelogram. Special products named and estimated indicated in their proportion to each other and to the total.

Silver $40,000,000

Hens' eggs $80,000,000

Wool $64,000,000

Pig-iron $90,000,000

Poultry and eggs . . . $180,000,000

Milk, butter, and cheese . $350,000,000

APPENDIX VI.

No treatise upon wages could be considered in any measure complete, without some reference being made to the great variation in the purchasing power of money. With wages at the same or nearly the same rate in the same place, one family will thrive upon an income on which another will almost starve. The reasons are not far to seek, but in order that the case may be fully comprehended, attention should first be given to the excellent and varied subsistence which may be procured at an apparently very small cost.

To that end I will first submit an analysis of the cost of food in a large factory boarding-house which is maintained by Messrs. Wm. E. Hooper & Sons, owners of some of the best cotton-mills in Maryland. This house was built to meet the wants of many women who came to work in the village where they had no relatives, and who were compelled to board in insufficient quarters, sometimes four in a room, or were in other ways subjected to injurious conditions.

As such statements as this possess a permanent value, being very difficult to obtain in a reliable form, I will give the cost in all its detail of the food of these adult women for six months.

EXPENSE ACCOUNT, JAN. 1ST TO JULY 1ST, 1884.

GROCERIES.		AVERAGE PRICE.	
Flour	30 bbls.	$5.40	$162.00
Corn Meal . .	245 lbs.	.05	12.25
Buckwheat . .	1 "	.05	.05
Rice . . .	80 "	.06¼	5.00
Hominy . . .	2¾ bus.	1.40	3.85
Crackers . . .	33¾ lbs.	.08	2.70

GROCERIES.		AVERAGE PRICE.	
Sugar	2,291 lbs.	0.07\frac{36}{100}$	$168.74
Syrup	69$\frac{3}{10}$ gals.	.30	20.79
Teas	91 lbs.	.43	39.13
Coffee	540 "	.12$\frac{1}{2}$	67.50
Yeast Powder	116 bottles	.12	13.92
Candles	32 lbs.	.12	3.84
Soap	1,074 "	.07	75.18
Soda (bicarbonate and washing)	116$\frac{2}{5}$ "	.01$\frac{1}{2}$	1.75
Allspice and Cloves	1 "	.28	.28
Nutmeg	1 "	1.00	1.00
Mace	—	—	.15
Ginger	2 lbs.	.12$\frac{1}{2}$	.25
Pepper	15$\frac{1}{4}$ "	.17$\frac{1}{2}$	2.67
Mustard	14$\frac{4}{5}$ "	.25	3.70
Cinnamon	$\frac{5}{17}$ "	.34	.10
Flavoring Extracts	12 bottles.	.12$\frac{1}{2}$	1.50
Hops	4 lbs.	.35	1.40
Matches	$\frac{4}{5}$ gross.	2.50	2.00
Indigo Blue	$\frac{2}{5}$ "	2.50	1.00
Salt	1$\frac{3}{4}$ sack.	1.50	2.62
Vinegar	32$\frac{1}{2}$ gals.	.20	6.50
Saur-kraut	1 bbl.	—	11.50
Starch	68$\frac{1}{4}$ lbs.	.04$\frac{1}{2}$	3.07
			614.44
VEGETABLES.			
Potatoes	77$\frac{1}{8}$ bus.	.49	37.75
Corn (in cans)	33 doz.	.95	31.35
Tomatoes (in cans)	24$\frac{3}{4}$ "	.85	21.04
Beans	20$\frac{3}{5}$ bus.	1.25	25.75
Peas	8$\frac{1}{5}$ "	1.25	10.25
Turnips	1 "	.35	.35
Parsnips	3 "	.63$\frac{1}{4}$	1.90
Cabbage	307 head.	.07$\frac{3}{4}$	23.79
Onions	17 doz. bunches.	.25	4.25
Radishes	253 " "	.01$\frac{3}{4}$	4.43
Lettuce	330 head.	.02	6.60
Rhubarb	109 bdls.	.04$\frac{1}{4}$	4.63
Beets	116 "	.04$\frac{1}{4}$	4.93
Cucumbers	9$\frac{1}{4}$ boxes.	.75	6.93
Cymblings	3 "	1.12	3.36
Carrots	25 bdls.	.03	.75
			$188.06
FRUITS.			
Apples	1 bu.	.63	.63

Fruits.		Average Price.	
Berries	186 boxes	7c.	$13.02
Currants	11½ lbs.	9	1.04
Raisins	11 "	10	1.10
Prunes	16¼ "	7½	12.12
Fruit Butter	68 "	7	4.76
			$32.67

Meats.			
Salt Meat, Ham	652 lbs.	13c. to 15c. per lb.	
" " Shoulder	626 "	8 " 11 "	209.40
" " Breast	288 "	10 " 11 "	
Beef, Roast	1,034 "	10 "	
" Steak	1,360 "	12½ "	338.43
" Soup	137 "	9 "	
" Corned	527 "	10 "	
Pork	213 "	11c. and 12c. per lb.	24.66
Lamb	97 "	12 "	11.64
Sausage and Pudding	446 "	10 "	44.60
Liver	137 "	8 "	10.96
Scrabbles	22 "	15 "	3.30
Tripe	28 "	8 "	2.24
Lard	434 "	10 "	43.40
	6,001		$688.63

Oysters	4 gals.	$1.11	4.44
Fish			26.42
			$30.86

Butter	462 lbs.	20c.	92.40
Cheese	69 "	15	10.35
Eggs	264 doz.	16	42.24
Milk	473 gals.	24	113.52
Mince Meats	155 lbs.	10½	16.27
			$274.78

RECAPITULATION.

Groceries	$614.44
Vegetables	188.06
Fruits	32.67
Meats	688.63
Oysters and Fish	30.86
Butter, Cheese, etc.	274.78
	$1,829.44

Fifty-nine women were boarded six months, an average of $26\frac{1}{4}$ days each month, which gives 9,292 days' board. The cost of food was \$1,829.44, or at the rate of $19\frac{68}{100}$ cents per day for each boarder.

The exact number of servants is not given with this short term, but is given with the following statement, covering four years, 1880 and 1883 inclusive. If the proportionate number of servants be added for the six months covered by the foregoing details, it would doubtless reduce the cost of food per capita to about 18 cents as against 20 cents for the previous four years, giving an example of the general reduction in the cost of subsistence which has occurred in the year 1884.

Without going into the exact details, the cost of conducting this boarding-house for the four years, 1880 to 1883, will next be given, and the proportions of the food will be shown by the graphical method. (See page 162.)

It will be observed that the number of days represented by the boarders is 99,456
To which must be added for the servants   17,520

                Total   116,976

which total being divided into the cost of the food, gives a result of a fraction less than *twenty cents* per day.

Many curious points will be observed in this bill of fare.

1st. Nearly every one will be surprised at the relative cost of sugar, as compared to farinaceous food. This case is not exceptional,—such is a very common almost universal rule.

2d. The very small use of corn meal as compared to wheat flour. The use of corn meal as the principal farinaceous food appears to be confined to the black population of the South; next to them the Yankee of New England makes the greatest use of " brown bread " and " Johnny cake." It is also apparent that two very important and nutritious articles of New England diet are wanting in Maryland, to wit : cod-fish-balls and baked beans.

3d. The quantity and variety of vegetable food.

Food, Fuel, and Light.

Meat . .	$8,866	Fresh beef, $3,656
Dairy and eggs .	4,490	Butter, $2,761
Vegetables . .	2,800	Potatoes, $1,103
Flour and meal .	2,444	Flour, $2,278
Sugar and syrup .	2,311	
Tea and coffee .	1,011	T | C
Fruit . . .	592	
Spice, salt, etc. .	537	
Food .	$23,051	

116,976 days. Cost of Food only, 20 cents per day.

Food, Fuel, and Light, 22½ cents per day.

} This ratio is probably one third above the average.

Fuel 2,562

Light 706

Food, F., and L. $26,319

4th. The large proportion which fresh beef bears to all other meat.

Consideration may next be given to the cost of subsisting prisoners in all the jails of Massachusetts.

These prisoners are served with the best quality of bread; beef which consists of the carcases of beeves of first quality, from which the best cuts have been taken for hotels, the remainder of such special stock being purchased on contract; vegetables, tea, rye coffee, sugar, and other articles substantially necessary.

The average cost of the materials used for food delivered at the jails in 1883 was $44.45 per head, or a trifle over 15 cents per day for each prisoner.

But the subsistence of the employés in the prisons is included in this sum, and they constitute over ten per cent. of the whole number whose food is represented in this cost-statement, while their food is doubtless more varied. This reduces the average, and in some of the larger jails the economy of material is greater, so that the cost per head is even as low as 12 cents a day.

In the separate prison for women each prisoner is weighed when committed and when discharged, and almost all gain in weight during their term of imprisonment.

The cost of food in this women's prison for 1881 was $14,713.04, which sufficed for the supply of prisoners, employés, and officials for 98,550 days, or a fraction less than 15 cents per day.

Next we may consider the cost of subsisting factory operatives in New England,—male and female.

I have been able to obtain only one statement from a village in Central Massachusetts, as follows:

COST OF BOARDING 17 ADULT MEN AND 8 WOMEN (3 SERVANTS) FOR SIX MONTHS, IN 1884.

Meat and fish	$540
Butter, cheese, eggs, and milk	336
Vegetables	72
Flour and meal	132
Sugar and syrup	87
Tea and coffee	54
Fruit, green and dry	33
Spices and salt	24
Total	$1,278

This sum represents 4,575 days' board at 28 cents per day to each boarder. These boarders being principally men engaged in arduous mechanical work, it will be observed that the quantity of food, especially of meat, is large, and the cost correspondingly high, as compared to the subsistence of women in Maryland.[1]

On the basis of these and other data which have come under my notice, there can be no question that an ample and varied supply of nutritious food can be supplied in the Eastern portion of the United States at a cost not exceeding 20 cents per day, or $1.40 per week, and probably for a less sum in the West, provided it is judiciously purchased and economically served.

I have given the cost of the rations of "hog and hominy," *i. e.*, bacon and corn meal, furnished negro laborers at the South at a cost of 50 to 70 cents per week, to which must be added chickens raised by themselves (or by others), vegetables (each laborer customarily having a garden patch), fish which abound in many places, sugar, molasses, and salt. Perhaps $1 a week would cover the whole.

If it is suitable to assume that these three classes, to wit :

1st. Adult women engaged in factory work in Maryland ;

2d. Prisoners in Massachusetts jails, mostly adult men ;

3d. Workmen and factory operatives, male and female, in New England, may be taken as exponents of the consumption of food necessary to comfortable subsistence throughout the country at an average of 20 cents per day, or $73 per year, then the total cost of necessary food of the population of the United States, in the census year, might be approximated as follows :

---

[1] A very large portion of the students in Harvard University take their meals at a "commons" table in Memorial Hall which is conducted by an efficient steward, and the actual cost is divided per capita. During the terms of 1883-4 the average cost of food per week was $2.59, or 37 cents per day. Preparation and service, including steward's salary, brought the charge to each student up to $4.12. The cost of the first month of the autumn term of 1884 has been reduced to $3.97 for the whole service.

Persons 10 years and above, numbering 36,761,607, at $73 per year $2,683,597,311
Persons below 10 years, numbering 13,394,176, at $40 per year 535,767,040

    Total $3,219,364,351

Even this sum would be more than twice the value of all clothing made from textile fabrics, domestic and foreign, carpets, upholstery fabrics, laces, ribbons, etc., etc. ; with garments, included, also including buttons, tapes, and other materials used in garments ; and also including the first washing, starching, and packing of shirts or other similar garments when made in factories, which items in the case of shirts, cost more than the making or stitching ; all of which I have computed at the lump sum of $1,500,000,000.[1]

But while textile fabrics and garments of all staple or necessary kinds are sold at the least possible margin of profit ; and while every scrap of waste is saved ; also while garments, as a rule, are worn out by some one before they go to the paper-maker or to the shoddy mill to be reconverted,—most articles of food are subjected to the greatest waste, either in purchasing, cooking, or in consumption.

The examples which I have taken represent food purchased in considerable quantities at wholesale prices, cooked properly and with economy, and used carefully with the least measure of waste. Yet at this average the value of the food would be $3,220,000,000
Add the clothing and other textiles . . 1,500,000,000

   Making a total for food and clothing of . $4,720,000,000

---

[1] This lump sum was reached by taking as a basis the census value of all the textile fabrics made in the United States, adding thereto the imports, then sorting out those which were ready for consumption as they come from the factory. The remainder, being materials used in garments, were then computed as clothing, by obtaining the average ratio which the value of the cloth bears to the completed garments ready for sale. The result must be very nearly correct, and it gives an average of $30 per head of population for clothing, carpets, laces, ribbons, and other textiles.

which is a little less than fifty per cent. of the sum of my computation of the total product of the country, and is over fifty per cent. of the total consumption of the country aside from additions to capital, estimated at $9,000,000,000.

But at this ratio food would be only about thirty-three per cent. of the whole cost of living. Now it will be observed that Dr. Engel, of Berlin, Carroll D. Wright, of Mass., and other fully competent authorities compute the ratio of the prime cost of food consumed in the families of workingmen at fifty per cent. of their income in respect to the thrifty and well paid, but at sixty per cent. of the whole income of common laborers or persons whose wages are low. Therefore this low ratio is not a true one, and the actual price or cost of food is doubtless more than this standard and the difference between $3,220,000,000 and a sum perhaps fifty per cent. greater is the measure of the waste or want of economy in the purchase and use of food.

No one can doubt that the actual cost of food prepared for use in workingmen's families would be on the average either twenty-five to forty per cent. more than the standard of twenty cents a day in money, in the more densely populated parts of the country; or else, if only twenty cents a day were spent, it would fail to yield half as good a subsistence as is obtained in the establishments cited, for want of skill both in purchasing and in cooking.

Let it be observed that while it is proved by these statements that an ample and varied subsistence *can* be supplied to adults at twenty cents a day, even in the Eastern states which are most distant from the fertile plains of the West, no such economy is realized except under similar conditions to those cited. But if we add only five cents a day on the basis of the census population we must add $912,500,000 to the aggregate cost, and at ten cents more we must add $1,825,000,000.

The former sum, added to the previous computation of $3,200,-000,000 at twenty cents a day, would bring the total cost of food at the place of consumption up to $4,100,000,000, or twenty-five cents a day, which would be still far less than fifty per cent. of the commercial product of $9,000,000,000.

Now these figures may be true or they may be merely visionary statistics, but they correspond fairly well with the deductions of economists who have examined into the conditions of particular families. Even if we add $1,000,000,000 to the computation of 20 cents a day we still fail to reach 50 per cent. of our computed commercial product.

Whether accurate or visionary these computations bring out one great fact in the clearest manner, namely, that the greatest cause of want in this country is waste. Whoever can teach the masses of the people how to get five cents worth a day more comfort or force out of the food which each one consumes, will add to their productive power what would be equal to one thousand million dollars a year in value.

How can this be done? In my treatise on "The Railway and the Farmer" I have given a diagram of the cost of bread in New York, showing it to be less than three cents a pound, and I have shown that it can be profitably sold at half a cent per pound profit, or at six cents for a loaf weighing $1\frac{3}{4}$ pounds, if the sales are made on a large scale over the counter for cash.

But the price of bread in Boston in the small shops is five to eight cents a pound.

Fish, meat, vegetables, and fuel, when sold in small quantities are subject to as great or a greater advance on the first cost. The grave difficulty is to cheapen the distribution of perishable commodities. There is no such difficulty in regard to textile fabrics, flour, sugar, or other staple articles.

In the body of my treatise I have made the statement that the highest rents are paid in cities for the right to make use of the warehouses or shops in which the largest amount of goods can be sold at the least possible profit or advance on the first cost. This rule applies to every branch of wholesale distribution, and also applies to the retail distribution of staple dry goods as well as of flour, of sugar, and of a very few other articles of food ; but it seems to have no application to the retail distribution of meat, vegetables, fruit or to the conduct of any of the small shops in the poorest districts. Far be it from the writer to impute blame or fault to the small shopkeepers, bakers, or grocers who supply the very poor. Dealing in small quantities, often granting dangerous credits, and paying rents which are relatively very high in ratio to the amount of their possible traffic, their small gains necessarily constitute a large ratio, or per cent. on each article sold. In this as in other matters, systematic organization, the use of a large capital and the custom of making very large sales at very small profits must justify the great traders who have absorbed so many small establishments. Again, we must revert to relative proportions. Out of 17,392,099 persons engaged in all kinds of gainful occupation in the census year there were only 1,810,256 occupied in trade or transportation, or between 10 and 11 per cent.; but in just the measure that this force can be reduced will the cost of distribution be lessened.

One may well study the methods of one of our great retail dry-goods stores or shops as an example of what might perhaps be accomplished in the distribution of food in the same cities.

Dry goods, so called, of all staple kinds are distributed at a very small advance on the wholesale prices, and it is doubtful if any organization could be invented for lessening the cost below what it now is. The chief profit of the dealers, as well as the principal customs revenue of the Government, is derived from goods which depend on their style and adaptation to the passing fashion of the season,—or from laces, ribbons, and small wares, while staple and useful goods are sold at a fraction above their cost.

It is, of course, vastly more difficult to systematize the distribution of perishable commodities, but perhaps it may be done.

This is the great problem of city life. How shall the rate of wages, whatever that rate may be, be made adequate to the wants of him who earns it?

Let it be remembered that this rate is the measure of the laborer's share of all there is produced, but that all there is is IN EXCESS *of all the wants of our whole population. The rate would suffice for an ample subsistence for every man, woman, and child in all our broad land, if only the mechanism and the metaphysics of distribution could be brought within the rules of social science.*

Cannot bread be served to the workmen of Boston at three cents a pound, as well as in New York or in London?

Cannot the waste heat of the bread ovens be used to stew meats and to make strong beef broth to be sold over the counter with the bread?

Cannot methods be adopted for bringing milk and vegetables within easier reach of the poor, who need them most?

Cannot as good a subsistence be supplied outside the prisons at 12 cents a day as can be furnished within their walls?

In other words, must an honest man become a thief and be sent to jail in order that an ample supply of excellent food may be brought to his door at a cost of 12 to 15 cents a day, or one dollar per week?

The average which I have given is above the limit of a laborer's ration being 1\frac{40}{100}$ per week.

Such an average as this for the cost of an ample and varied supply of food will appear very small to most of the readers of this book, but it is not for such persons that much consideration is needed. The case to be provided for is that of the *common laborer* in a crowded city, the measure of whose share of the annual product is what one dollar a day will buy,—or perhaps one dollar and a quarter,—and upon whose work an average family of four other persons may depend, making five in all. His week's wages, assuming that he is in constant employment at $1.25 per day, will be $7.50. If it be assumed that the five members of his family consume the rations of three and a half adults only, then at $1.40 per week, the cost of food would be $4.90, leaving only $2.60 for rent, clothing, and other necessities of life, Of course such a proportionate expenditure for food is hardly to be considered, yet, upon the average determined by the investigations of Dr. Engel and Carroll D. Wright, such a man would expend sixty per cent. of his wages, or $4.50 per week for food. But then comes the question: How much food does he get for his money and how is it cooked after he buys it? On the answers to these latter questions rest comparative want or welfare. If common laborers in cities could be supplied with food as well and as cheaply as the prisoners in our jails, *i. e.*, at $1 per week, then in the case which I have assumed food would cost but $3.00, and $4.50 would be left for other expenses.

Cannot the distribution of meat, bread, fish, vegetables, and milk be organized and made profitable with large sales at small profits, as well as the distribution of calicos, blankets, and petticoats? Perhaps with a little more risk and a somewhat larger ratio of advance on cost because of their perishable nature, but yet in such a way as to reduce the present cost of subsistence in a very large measure?

Lastly, can cooking be taught in the public schools or elsewhere?

Cannot a waste of food equal to five cents a day on the average be prevented? Is there such a waste? If there is, its measure is over one thousand million dollars a year. Let him who doubts such waste glance at the contents of the dinner-pail of the next laborer whom he passes at the noon hour, or take a meal with an average laborer's family.

Upon the answers which may be given to these questions, the adequacy of the wages of workmen in cities will mainly depend, whatever the rate of their wages in money may be.

This is but another phase of the question which forms the title of this treatise:

WHAT MAKES THE RATE OF WAGES?

APPENDIX VII.

It may be suitable to assume that the average quantity of food served to adult women in a factory boarding-house in Maryland is a fair standard of the average consumption of the working people of the United States, in quantity if not in kind, and, if we apply the ascertained facts in this example by computations covering the whole population of the census year, numbering in round figures 50,000,000, we may reach an approximate estimate of the total value of food at the point of consumption, and then by comparison with the estimates of the value of farm products at the place of production, made in the Department of Agriculture, we may approximately test the accuracy of all the conclusions or hypotheses relating to the cost of subsistence, which are made use of in this essay.

The diagram given on a previous page gives the actual cost of the food consumed by seventy-four boarders and six servants, in the four years 1880, 1881, 1882, 1883, averaging twenty cents per day to each person, in this boarding-house.

It will be observed that this food was well bought, in considerable quantities at a time, and was economically cooked and served.

In the following table the proportion of each kind of food consumed by one person in one year is given in the first column, and in the second column the gross sum is given which this would represent, if each person in a population of fifty million enjoyed the same rations.

Article of Food.	Per Person.	For 50,000,000.
Meats (including poultry, fish, and oysters)	$27.70	$1,385,000,000
Butter, cheese, and milk	12.18	609,000,000
Eggs	1.85	92,500,000
Vegetables	8.75	437,500,000
Flour and meal	7.64	382,500,000

Article of Food.	Per Person.	For 50,000,000.
Sugar and syrup	$7.22	$361,000,000
Tea and coffee	3.16	158,000,000
Fruit, green and dry	1.85	92,500,000
Salt, spices, vinegar, etc.	1.67	83,500,000
	$72.02	$3,601,500,000
Imported,—tea, coffee, most of the sugar, part of the fruit and spices, etc.	10.02	501,000,000
Product of domestic agriculture	$62.00	$3,100,500,000

It is easily proved that the consumption of sugar and syrup by these women was excessive in proportion to flour, but their consumption of meat was less than that of adult men in Massachusetts, as will presently appear; while children under ten years of age would consume less than either, and the very poor or the common laborers of the country would be able to buy less meat and sugar, and would depend more on grain and fish.

All that is assumed in the comparison which follows, of the foregoing total with the computed value of the food products of agriculture made by Mr. J. R. Dodge, is that the aggregate consumption of eighty working women in Maryland is a fair standard by which to measure a good and sufficient proportion of food for the whole population.

Before we venture upon this comparison, we may observe the cost of the larger ration of each one of the seventeen adult men and eight women in Massachusetts, at the rate of 28 cents a day, or $102 24/100 per year. This annual ration, when subdivided, was as follows, for one person one year or one day:

	Per Year.	Per Day.
Meat and fish	$43.20	.1182 cents.
Milk, butter, cheese, and eggs	26.88	.0737 "
Vegetables	5.76	.0158 "
Flour and meal	10.56	.0290 "
Sugar and syrup	6.96	.0190 "
Tea and coffee	4.32	.0118 "
Fruits—green and dry	2.64	.0073 "
Salt, spice, vinegar, etc.	1.92	.0052 "
	$102.24	28 cents.
Imported about	11.24	
Domestic production	$91.00	

If we consider this average expenditure per day, we may find that, although its measure is without doubt a large one as compared to the expenditure for food which is possible to the common laborer, especially in cities, and that, even though such a laborer should spend as much money he could not get as much for it—yet, even when the money is well spent, it will not give any excessive quantity, or very extra quality of provisions.

The laborer who should spend 28 cents per day for food in the most intelligent manner, to be prepared at home in the proportions indicated, could obtain in Boston to-day for $11\frac{82}{100}$ cents, about half a pound of good beef, mutton, or poultry; about three quarters òf a pound of fair quality of meat; or one pound of coarse fresh meat, salt meat, sausages, or fresh fish.

$7\frac{37}{100}$ cents spent on dairy products and eggs would give him half a pint of milk, 2 oz. of fair butter or 1½ oz. of good butter, 1 egg—"shop 'un,"—and a scrap of cheese.

$2\frac{90}{100}$ cents would give him half a pound of good bread, of meal and flour equivalent to about one pound.

$1\frac{90}{100}$ cents would give him between 3 and 4 oz. of sugar; and $1\frac{18}{100}$ cents spent on tea and coffee might give him one cup of each per day, or one of either, night and morning.

This would be the utmost if the money were spent with care and intelligence; whether the money would yield 50 or 75 per cent. as much would depend upon the personal capacity of him who spent it. The average laborer would probably obtain about as much for 28 cents as the Maryland factory operative enjoys for 20 cents per day, the food of the latter being well bought.

These data are entirely insufficient as a basis for rules; they are merely given as an indication of what might be accomplished in improving the distribution of food if the Chiefs of the Bureaux of Statistics of the several States would adopt a uniform schedule and plan for ascertaining the relative proportions and cost of the food consumed in the private families of workingmen and women. When the facts are known the method of improvement may become apparent.

As this average ration is used as a standard, then, counting each two children of ten years or less as one adult, the total consumption of food of domestic production in the census year would have been valued at about $3,890,000,000,—but it is hardly to be assumed that the average adult person enjoyed as large a ration as did these men and women in New England, whose food was carefully purchased and served. The average of the Maryland women is probably a much truer standard of comparison.

In each case attention is called to the vast aggregate of the value of dairy products and eggs, indicated by both these tables, and to the fact that the money-cost of sugar and syrup is not less than eighty per cent. of the cost of the flour and meal. Several years since, when the writer had the direct charge of a large cotton factory, when sugar was much higher in price, he found that the sugar consumed by a large body of French Canadian operatives cost more than the flour.

Counting two children as one adult, and then extending the ration of butter, cheese, milk, and eggs in the Massachusetts boarding-house to the whole population, the aggregate of this one item would have been over $1,000,000,000, at retail prices.

These statements may seem to possess only a curious interest; but may it not be held that, when special legislation is demanded in order to sustain special interests—as in the case of the silver product,—some standard of comparison should be established by means of which the utter insignificance of the silver production of $40,000,000 may be made apparent? These are retail prices, and it is just at this point of final or retail distribution that debasement of the currency works the most malignant fraud.

One needs only to recall the manner in which shrewd buyers availed themselves of the opportunity to buy great stocks of goods when the legal-tender notes issued during the war began to depreciate, and then availed themselves of the rise in prices which ensued to make huge fortunes, to comprehend the result which will follow the depreciation of our present currency when the light-weight silver dollar, worth only 85 cents, drives gold from circulation. But in the present case this malignant effect will be rendered more intense, because there will be no war demand to stimulate production. Constructive and productive enterprise will be reduced to the point of absolute necessity, and the rate of wages will be thereby reduced at the very time when the money in which wages are paid will lose fifteen per cent. of its purchasing-power.

Such are the consequences of fraud perpetrated under the forms of law, and such will be the consequences of the continued coinage of silver dollars, if members of Congress continue to submit to the dictation of the so-called Silver States, whose whole annual product of silver is worth less than half the annual product of hens' eggs.

If reference be now made to the estimate of Mr. Dodge, of the value of agricultural products at the points of sale nearest the farms, which is what I understand to be the "farm value" so-called, we find the product of grain, meat, dairy products, vegetables, and other articles which are used for the food of men, estimated at $2,900,000,000; but from this estimate corn fed to beasts of burden should be deducted, and hay, converted into meat and dairy products, should be added. It may be assumed that one would about balance the other, and that the net value of human food at the farms was as above stated.

In the essay on "The Railroad, the Farmer, and the Public," I have computed the proportion of food products moved by rail at one half the total tonnage. On this basis we must add $200,000,000 for the cost of moving food from producer to consumer.

Net value of food at the farms	$2,900,000,000
Cost of transportation	200,000,000
	3,100,000,000
Deduct exports estimated	440,000,000
Remainder	$2,660,000,000

By comparing this sum with the previous computation of the value of food consumed, at the place of consumption, we find a

difference of $440,000,000, which is easily accounted for as follows :
1. Cost of milling grain and barrelling flour.
2. Cost of slaughtering and packing animals.
3. Admitted underestimate in the computation of Mr. Dodge, in respect to the production of vegetables and orchard products.
4. Cost of distribution at wholesale.

So far as the data can be obtained, the difference between the two sums would be fairly covered by these items, and the computation of consumption therefore fairly sustains the estimate of production at farm values if the standard adopted is a true one.

Attention must however again be called to the fact that the food purchased for the Maryland boarding-house was bought in considerable quantities, and made to serve its utmost purpose; therefore, a considerable addition must be made for the cost of more luxurious consumption of the more prosperous classes, and it must be borne in mind that the small purchases at retail for single families will give each person a much less quantity of food for the money spent, or the same money spent will buy a less quantity of food. Hence it may be fairly assumed that the proportion of the productions of agriculture consumed in the country, which bore a value in the census year of $2,660,000,000 at the farms, finally cost the consumers at the point of consumption about $4,000,000,000 or $4,500,000,000; which sum would represent an average of $80 to $90 per year, or a fraction less than 22 to 25 cents per day, for each person of a population of 50,000,000. In saving a part of this vast difference which doubtless exists between the farm value of food, $2,600,000,000 (with the charge of $200,000,000 for transportation added thereto), and the sum paid at retail for the same quantity, is to be found the greatest opportunity for economy and for benefit to common laborers, especially in crowded cities.

In this computation I have paid no attention to the conversion of grain and fruit into whiskey, beer, and wine, as I know of no accurate method of computing the excessive cost of distributing liquor by the glass. In view of the prime cost of whiskey and beer, coupled with the fact that a revenue of about $90,000,000 is derived from the excise tax upon them; and also bearing in mind the ratio which the price of a glass of liquor bears to the cost of a cask, it may be safely assumed that drink bears a ratio of ten per cent to the cost of food, or about $400,000,000.

Upon the basis of these computations, food, drink, tobacco, domestic fuel, and light cost consumers $4,500,000,000 to $5,000,000,000; clothing, carpets, and other textiles as previously computed, $1,500,000,000. Total, $6,000,000,000 to $6,500,000,000, or $120 to $130 per year to each person, on the basis of the population of the census year, leaving $4,000,000,000 to $4,500,000,000 for all other expenses of living and for profits, on the basis of a total of $10,000,000,000 to $10,500,000,000 product.

In submitting this final analysis of so complex a problem it might be prudent for the writer to add the customary *caveat*, which would be consistent with his long practice as an accountant, " E. and O. E."—*i. e.*, " Errors and Omissions Excepted." [1]

---

[1] A second edition of this essay may be called for. Readers who are in possession of statements of the cost of food corresponding to the one herein given from the books of the Maryland factory boarding-house, will confer a great favor on the author if they will send them to him. Address P. O. Box 112, Boston.

CONCLUSION.

If the principle which is submitted in this treatise can be sustained, to wit : that by the competition of capital with capital the annual product is increased while the relative share of the capitalist is diminished, and that no more can possibly be saved and added to the capital of a given country in a normal year, or series of years of peace and order, than is necessary to keep land, buildings, machinery, and tools in a condition of maximum efficiency ; if it be also true that by the competition of labor with labor, aided by capital, the aggregate of products is increased, of which aggregate the laborers receive a constantly although slowly increasing share, both absolutely and relatively,—then it follows that progress and poverty have no natural or necessary relation to each other under existing customs, or as a consequence of competition.

If it be also proven that the measure of all there is produced in a given year, when converted into terms of money by bargain and sale—in other words by exchange—must be the source of all profits and wages, and that whatever this sum may be, it constitutes the limit beyond which profits and wages cannot go, then it also follows of necessity that by so much as one man secures more, may some other man have less of what has been produced in that year.

But it by no means follows that the welfare of the one is the cause of the want of the other ; there is enough for all, and the common cause of want is usually ignorance, unwillingness, or incapacity to do the kind of work which is waiting to be done. It would not be a pleasant thought to any man to feel that his larger share of what there is has been attained at the cost of his fellow-man, and such is not the fact.

How then shall the just man justify himself if he be rich and prosperous and if his family each consume far more than what 40 or 50 cents a day will buy if that be the average share ; or if each consume more than his average measure of all that is produced, whatever it may be?

The only answer to this question is that he must either work with brain or hand in such a way, or make such use of the capital of which his wealth in part consists, that the general production shall be increased in greater proportion by means of his work than the measure of his own consumption. And this is the exact function of the capitalist. In one sense he employs labor —but it is quite as true that labor employs him. Just as instinctively as an army of soldiers recognizes its true leader, does an army of laborers choose its own capital. The best workmen select the best mill ; the best managers are always chosen by the best workmen to serve them and to be served by them.

When capital under skilful direction doubles the productive power of each laborer, and leaves him the larger part of the increase, personal wealth and common welfare become synonymous terms; while, on the other hand, he who wastes but does not increase production in any way, however rich he may be, is really but a pauper—that is a person who is supported at the public cost.

There is something very merciless in these figures which make the rate of wages, and in the face of them the shallow nostrums of the greenback party, and of the common ruck of so-called "labor reformers" who infest the lobbies of the legislature with all sorts of empirical projects of law, become worse than an impertinence.

It is true that the measure in money of all there is produced and commercially distributed in this country may vary a little from fifty cents' worth per day to each person, including all profits as well as wages ; or forty cents without profits ; it may be a little more, it may be a little less.

The measure of the savings or increase of capital may vary

slightly from five cents a day per capita, which is the proportion that I have set aside as the probable amount.

The measure of all the taxes which are now three and a quarter cents a day to each person can be reduced to two and three quarter cents and no more.

Whatever the true averages may be, each of these variations of a cent or less would count in millions.

One cent a day added to the resources of all the people, or three cents a day added to the average wages or earnings of those who do the work, renders an increase of the national product necessary, of over two hundred million dollars' worth a year and a market must be found for the increase.

One cent a day taken from wages and added to savings would alter the computed sum of the annual addition to capital from one thousand to twelve hundred million dollars, or twenty per cent.

Half a cent a day remitted from our excessive taxation would take off one seventh of the whole burden, amounting to one hundred million dollars a year.

On such fractions as these prosperity or adversity depend.

A margin of only a cent or two a day to each person is all that separates national want from national welfare, or rather it indicates the difference in conditions, because on such a margin of profit or loss on the whole traffic of the country constructive activity or weary depression may be determined.

If our population, January 1, 1885, shall be 58,000,000, two cents a day profit on each person's consumption would be $423,400,000—a sum of profit which would set every wheel of industry into most rapid motion. Two cents a day loss would bankrupt thousands of merchants and stop more mills and works than are even now idle.

When legislators pass acts by means of which they intend or expect to control the course of productive industry and to raise the general rate of wages, they may well ask themselves how they can add one thousand million dollars' worth to our pres-

but over-abundant production, find a market for the increase, and so regulate the distribution of the proceeds of the sale as to give each person five cents a day, or each working man or woman fifteen cents a day, more than they now enjoy.

As well might each member of Congress try to add one cubit to his stature as to attempt to do this thing—but if members of Congress cannot construct they can obstruct. They can divert the wage fund of the many to the profit of the few by acts of taxation for the support of private interests, as in the purchase of silver.

At any moment the rate of wages of every man, woman, and child now working in the employment of others may be impaired by many cents a day, if the coinage of silver dollars of light weight and worth only a little over 80 cents is not stopped. The tax alone which is imposed in order to buy the silver is but a trifle, but the malignant effect of tampering with the standard of value is the worst evil that legislators can inflict upon the people.

The completion of the reading of the proofs of the principal part of this treatise happens to fall upon the fourth day of November, 1884. Two days from this time it will be announced that one or the other candidate for President has been chosen to govern the United States for the ensuing four years, together with a Congress upon whom the duties of legislation will fall. After it shall have been announced " straightway all the people will return to their usual occupations, and will govern themselves according to their common habit." But their material welfare may be greatly affected by the measures upon which this or some other Congress must soon act. This Congress will have been chosen with little or no regard to the convictions of its members —if they have any convictions—upon the great fiscal questions which must come before it, and after no adequate discussion upon them among the people, yet it will find itself compelled to grapple with the problem of the safe method of abating taxation ; it must deal with the coinage of silver ; it must face a necessary change in the national banking system which will ensue under the rapid payment of debt. It must deal with financial problems of greater difficulty than those in which the reputations of the greatest men of England have been made or lost. With each and all of these great problems it will probably deal in a purely empirical and inconsistent manner for want of adequate leadership, and without such party responsibility as is necessary to the right conduct of representative government. And as it may deal with them may confidence or distrust control events, and may wages and profits alike be left free to increase or be gravely diminished.

It will probably happen that such a Congress will accomplish little,—nothing,—or perhaps worse than nothing.

In this event, the election of the next Congress, which will be free from the personal issues that have degraded the present election, will proceed on the basis of a discussion of measures rather than of men.

When this fortunate period arrives, the true era of reconstruction, both North as well as South, will have opened.

Since the end of the Civil War only the crudest measures have been adopted for the reorganization of industry.

Equal suffrage has been established so far as it can be assured by national statutes.

The restoration of a specie standard has been brought about, but whether it shall be a true standard of gold coin or a false one of debased silver coin is not yet determined.

The abatement of some of the most onerous taxes has been accomplished, but yet more remains to be done, and the real test of political intelligence and of statesmanship is before us, in doing this necessary work.

Under what conditions, whether of apparent general prosperity or adversity in this and in other countries, the next Congress may be chosen, no one can predict.

It must be very clear, even to the most superficial observer, that the present conditions of an apparent excess of production and want of market, in all the nations which have applied machinery in the largest measure to the work, whatever their fiscal or financial policy may have been, must have been caused by forces over which statutes can have little or no effect.

May it not be that the complete revolution in the methods of commerce, which has been brought about during the last twenty years by the railway, the steamship, and the telegraph, as well as by the application of science and machinery to agriculture, has now come to what might be called a culminating point, after which the benefits heretofore enjoyed mainly by the producers and distributers of the staple products necessary to life are now in process of wide distribution among all consumers? In other words, may it not be that under the beneficent law of diminishing profits and increasing wages a lower plane of prices on a gold basis has been reached, which is of a permanent character? This change has, for the time being, disturbed all the existing relations of labor and capital, and is destroying many great fortunes, while bringing many to whom the struggle of life seemed to be ended again to the necessity of arduous work; but will not the end be greater abundance to the laborers who constitute the great mass of the people, shorter hours of labor, and less arduous conditions of life,—at least among English-speaking people, and especially in this country?

I have attempted to show how very large a proportion of the work of production and distribution of this or of any other country must go on, whether the "times" are "good" or "bad." I have endeavored to prove that the difference between "good" and "bad" times in a nation which is at peace, consists mainly in the question whether *one* per cent. of the population or *three* per cent. of the working force is idle, or whether there is work seeking to be done which would give employment to *one* to *three* per cent. more men than are to be found ready to do it. In other lands, and in former times in this country, the full employment of the people, or lack of employment, has been a question of abundance or scarcity; but in this country at the present time, as well as in 1873, no suspicion of scarcity has been suggested. An excess of production exists, and in the effort to get it into use, the rate of interest is reduced,—what is called "plenty of money" is seeking borrowers. This plethora of money is merely our excess of grain, timber, coal, iron, cotton, wool, and cloth, seeking consumers. The title to this excess is measured in terms of money, and is deposited in banks ; and banks, bankers, and trust companies seek to find consumers who will pay interest for its use. That is their function : they lend titles to consumable property or quick capital measured in terms of money, but the proportion of actual money used in these transactions is less than five per cent. of the aggregate. It is the excess of certain special products seeking to find a wider market that depresses the rate of interest. It is not money which is so plentiful, although there is enough of that ; it is quick capital, iron, coal, cotton, corn, wheat, oil, seeking consumption, the title to which is held in trust by banks and bankers, and for which they seek borrowers. When a wider market can be found for what we call over-production, not only will the rate of wages be maintained, but the rate of interest or profit on invested capital may also be enhanced.

This wider market need not be a foreign one,—we have still a continent to subdue, wanting only confidence and constructive enterprise.

One per cent. of our population, numbering over five hundred thousand workers, who sustain fifteen hundred thousand persons, may be waiting to use the iron, coal, and timber, to eat the grain and to wear the cloth, but cannot get it.

Coined money is plenty ; other nations send coin itself with which to buy a part of our excess ; but more of the excess remains, and yet the work of constructive consumption does not begin. It is the old nursery tale repeated—the pig won't go to the market, the dog won't bite the pig, the stick won't beat the dog, the fire won't burn the stick, the water won't quench the fire—and so on.

Why does not the pig go to market ? Whoever can answer that question will solve the puzzle of financial crises in times of peace and plenty.

One reason has been given by the late Walter Bagehot, in one of his essays, entitled " Physics and Politics," in the following paragraph.

After speaking of the imitative quality of men, he says :

" The grave part of mankind is quite as liable to these imitated beliefs as the frivolous part. The belief of the money-market, which is mainly composed of grave people, is as imitative as any belief. You will find one day every one enterprising, enthusiastic, vigorous, eager to buy, and eager to order ; in a week or so you will find almost the whole society depressed, anxious, and wanting to sell. If you examine the reasons for the activity, or for the inactivity, or for the change, you will hardly be able to trace them at all, and as far as you can trace them, they are of little force. In fact these opinions are not formed by reason, but by mimicry. Something happened that looked a little good, on which eager, sanguine men talked loudly, and common people caught the tone. A little while after, and when people were tired of talking this, something happened looking a little bad, on which the dismal, anxious people began, and all the rest followed their words."

There could be no more complete example of this imitative habit than may be found in the fluctuations in railway construction, which have occurred during the last twenty years in this country.

As soon as the war ended railway construction began ; it was pressed to the utmost ; every available man was set to work. "Something happened that looked a little good." Every sanguine railway promoter, honest or dishonest, " talked loudly, and common people caught the tone." Thus it went on until 1871, when over 400,000 men were employed in constructing over 7,000 miles of railway. Then something happened "looking a little bad, on which dismal, anxious people began to talk." The panic of 1873 occurred, and in 1875 railway construction had gone down to 1,700 miles, giving employment to less than 100,000 men. Next came the long struggle to resume specie payment, and in 1879 " something (the resumption of specie payment) happened extremely good." Every one became " enthusiastic, enterprising, vigorous, eager to buy." Railway construction was resumed, factories of all kinds were built, exports of farm products increased, sales were easy to make, consumption followed close on the heels of production. But there were many blunders. Useless parallel railways were promoted and built alongside of an adequate existing service, until, in 1882, 650,000 men were engaged in the construction of 11,500 miles, while over 450,000 men were engaged in operating existing lines. In 1882 one man in every ten of all who were occupied in any kind of gainful occupation aside from agriculture, was engaged in the construction or operation of a railroad.

"Something happened a little bad." Anxious men began to question the pace and to doubt the expediency of some of the work; all men took their tone; all stocks were affected, good and bad alike; construction fell off to not over 4,000 miles in 1884, and more than 400,000 men were discharged from work on this single occupation.

There were real causes or reasons for these changes, but their effect was exaggerated by over-confidence and too great distrust.

Almost every mile of the apparently excessive railway construction which culminated in 1872 has justified its use, if not its value to the original promoters, as almost every mile, with the exception of a few speculative parallel lines of the apparently excessive construction culminating in 1882, will yet be justified.

We now have 125,000 miles of railway, including as many or more through lines East and West, as can be profitably used for a long period, but the cross-way and connecting railroad service is totally inadequate, while many great States must double their mileage within a very few years. It cannot be long before this need will be felt; "something will happen that looks a little good"; confidence will return, and the ready and quick consumption of our excess will stop all talk about over-production.

Again, the construction of textile factories has wholly ceased. Yet if the construction of railways and other works were going on so that all workmen could afford to buy all the fabrics they need for themselves and their families, every spindle and every loom now in existence would be needed. Each branch of work waits upon the other ; one stops and blocks the way ; then another, then another ; but presently the "pig goes to market," and the whole procession moves on.

Each reason for a stop or a start is exaggerated by hope or by fear ; few have the instinct to foresee these tides of confidence and of distrust, but to them such tides lead on to fortune.

The habit of expecting greatly beneficial results from legislative action, or the reverse, is one of the principal causes of great fluctuations in the course of affairs, and from this habit of exaggerating the assumed power of legislative action in business matters comes the check to constructive enterprise when grave changes in legislation are pending.

Such changes have now become a matter of necessity and not of choice, and while they are pending every man waits the event and dares not plan for future enterprise.

Yet all the conditions of the country are ripe for prosperity. There is enough for all, and work is waiting to be done, that will soon become urgent, which would entitle every idle man or woman to an adequate share of the over-production which now clogs the centres of trade. The present Congress has proved itself incapable—it blocks the way. Before the next assembles, even the doubt which now causes stagnation may have yielded to the absolute necessity for constructive work to begin anew.

Whenever the new start, which cannot be long deferred, is made, the vast changes of twenty years, which I have faintly tried to picture, will be in full force, and the struggle for general comfort and welfare may then be less severe than it ever was before in this or any other land.

It has seemed to the writer that he could dimly perceive these beneficent results or promises amid the apparent confusion of the statistics, from which he has for many years endeavored to wrest their secret. True statistics are but the record of industrial history. He whose imagination cannot read what is written between their lines or interwoven in their columns, may rest content with the narrative of wars and dynasties, or of political changes, and may think he knows the true record of events; but can he tell how the people lived and moved, and how these wars and dynasties have been sustained. If he cannot, let him study what figures can teach to any one who knows how to master them, to wit: the industrial history of free nations. The battle is not always to the heaviest battalions, but to the people who can sustain the battalions longest. It is the commissary general who wins, without whom the master of the ordnance would be powerless. In the battle of life it is the same. If there were no prophecy of the future in this work, these computations would have no meaning, and the close study of the disclosures of the census would not be worth the time devoted to them.

I have ventured to call this a treatise upon "The Mechanism and the Metaphysics of Exchange."

The second term of this title may be most fully justified by a very slight consideration of the fleeting nature of capital: "All things have been others—all things will be others."

The term "fixed capital" when applied even to the most solid and substantial industrial works is yet a misnomer. There is absolutely nothing in the shape of productive capital which has any long duration among men.

The city warehouses, only thirty or fifty years old, fail to meet the need of modern commerce, except they be so completely reconstructed as to become almost wholly new. There is nothing left of the factory of fifty years ago, except a part of the foundation and the wheel-pit; and in that fifty years the whole of the machinery has been changed once, twice, or thrice.

The modern mechanic would scorn the tools which his father used, and would hardly be able to obtain a living by their use.

Of all the work which has been done by men to promote the exchange of services and the distribution of products, there is nothing permanent except the opening of the ways and the body of the laws.

It is safe to say that in one, or at the utmost two, generations every productive instrumentality now in use, except the opening of the ways, will be almost wholly without value because it will have been superseded by better mechanism ; therefore no treatise upon existing facts would be worth compiling if it did not give the greatest prominence to the metaphysical side of social science, and did not bring the imagination into play in forecasting the future.

The only bequest which one generation can give to the next must therefore be such development of the capacity of each individual as will enable him to grasp the opportunity that a free government may assure him, to work out his own material welfare by means of the mechanism which may be in use during the term of his own working life.

A true study of the Mechanism and of the Metaphysics of Exchange therefore is a true study of the History of Nations, and when a commercial history, even of the English-speaking people, is written in the way that it ought to be written, it will give us an insight into the one fact more important than all the rest, when it tells us how the masses of the people got their living—or, in other words what made the rate of their wages—amid the turmoil of wars, the contest of dynasties, the contention of creeds, and the struggle of the masses to overcome the privileges of the few when they had ceased to be founded on services rendered to the many.

For such a work as this might be, the compiler of this treatise can only prepare some of the materials affecting this country.

The recent publications of Prof. J. E. Thorold Rogers on "Work and Wages" for the last six hundred years, the "Growth of English Industry and Commerce" by Prof. W. Cunningham, and other English works, together with the investigations of Mr. Robert Giffen, cover a large portion of this ground in England.

EDWARD ATKINSON.

BROOKLINE, MASS., *Nov.* 4, 1884.

WHAT IS A BANK?
WHAT SERVICE DOES A BANK PERFORM?

A LECTURE GIVEN BEFORE THE FINANCE CLUB OF HARVARD UNIVERSITY,
MARCH, 1880

By EDWARD ATKINSON

NOTICE.—This tract is specially for the ACTIVE and COÖPERATING MEMBERS of the Society, and is not for general sale. If members desire any additional copies they will be furnished in any quantity at the rate of $10.00 a hundred, on application to the Secretary, or to G. P. Putnam's Sons, of New York, or Jansen, McClurg & Co., of Chicago, Ill., the publishing agents of the Society.

Respectfully,

R. L. DUGDALE,

Secretary.

BANKS AND BANKING.

A LECTURE DELIVERED BEFORE THE FINANCE CLUB OF HARVARD
UNIVERSITY, MARCH, 1880.

PUBLIC attention is very much devoted to the question of transportation. The importance of railroads and steamships is apparent to all, and every method that can be devised to promote the extension of their lines of traffic receives attention. From the dawn of history, commerce has been the measure of human progress. Upon the ancient caravan-routes of the Far East, over the Roman roads of a later period, across unknown seas, and by devious ways, commerce has from age to age extended its beneficent function. Even when nations have attempted to isolate themselves, by enacting excessive duties upon imports, the "fair trader," as the smuggler used to be called, has rendered the attempt of no avail. Men will exchange product for product, because there is no other way by which even a moderate degree of material welfare can be attained.

But in this, as in almost all branches of investigation, he who limits his thought or study to the purely physical side of the question will be misled.

In this apparently most material of all questions, how to subsist the human body, the work that is abstract or immaterial is of such essential consequence that railways, steamships, and canals would be shorn of more than onehalf their beneficent power if not rightly coördinated and worked in perfect harmony with instruments of distribution of a purely abstract, or perhaps we might say metaphysical, order. In this category come the operations of the bank and the banker.

But before we begin the discussion of the function of banks and bankers, of bills of exchange, bank-notes, and all other instruments by means of which the title to commodities is passed from one man to another, while the things themselves are being carried over the railway, it becomes necessary to give precision to our language, and to define the meaning of the words that we must use.

I am satisfied that a vast deal of bad legislation would be avoided if the graduates of high schools and colleges had more complete command of the English language, and more fully comprehended the exact meanings of common English words.

Before I can begin to consider the subject of banking, it first becomes necessary to define the word *money*. I shall assume that any young man who has had sufficient intelligence to pass the entrance examinations of Harvard University will know enough of the functions of money, and the qualities which it must possess in order that it may be entitled to the name, to warrant me in excluding stamped pieces of irredeemable paper, of late proposed to be issued by the government under the name of " fiat money," from the category of true or real money.

It is sometimes necessary, even for intelligent men, to consider the propositions in regard to what is called " fiat money," in order to prevent the uninstructed from being cheated by knaves, or misled by those whose intelligence on other subjects makes one hesitate to call them fools, but who must be classed among persons endowed with a kind of limited or perverted intelligence, for which the dictionary has not yet provided a suitable name. It is unfortunate that there should be even a few men among us whose influence has been established in the conflict with the slave power through which we have lately passed, who do not perceive the baleful character of the measures which they advocate. We could well spare them if they would migrate to the country which Boccaccio describes as the "Land of Mendacity," where they "use only paper money."

For the purpose of this lecture, without entering upon the history of money, I will limit the meaning of the word to the pieces of coined gold and silver used by most nations, under various names. Therefore, for the present, when I use the word "money," I shall mean gold or silver coins,— dollars, sovereigns, livres, francs, and the like. True money has been made of other substances in past ages, but at the present time nothing else is entitled to the name.

I am well aware that this limitation would not be admitted by many economists. It would be alleged that a law of the land makes the United States notes now in use "lawful money," as well as "legal tender," and that we must therefore accept the definition; but may not this very fact be cited as an example of the danger of corrupting the language?

If a word is perverted from its true meaning, it ceases to be an instrument of precise thought.

We have become so accustomed to the perversion of the term "money" from its strict application to the coined substance rightly so called, and its application to the promises of banks known as bank-notes, or to the promises of the nation known as legal-tender notes, as to make it difficult even to begin to speak to you on the subject of banking.

Another great and very mischievous perversion of the word "money" is to use it as synonymous with property.

We define a man's property by saying that he is worth a given sum of money, meaning only that his property would be measured by, or could be sold for, a certain sum.

It is from such perversions of the word that many men have been led to believe that welfare depends upon an abundance of money, and that "the times," as we say, are "easy" or "hard," just in proportion to the abundance or scarcity of money.

What is intended by the phrases "money scarce" and "money plenty," is more apt to be "capital scarce" and "capital plenty;" but there are also hard times when both money and capital are very plenty, and the real cause of adversity is that "confidence is scarce." We have lately passed through such a period.

One most potent cause of want of confidence is when the instrument used to serve as money is not true money; irredeemable notes forced into use by an act of legal tender are of this order. The more abundant such base or forged money, as may call it, becomes the less it serves its purpose. Depression, adversity, and loss, we have suffered in full measure during late years. Men have talked, with the wisdom of owls, of over-production, and have imputed the difficulty under which great masses suffered in recent years in procuring food, fuel, clothing, and shelter, to the alleged fact that we were over-producing corn and meat, that our mines delivered too much coal, that our looms wove too many yards of cloth, and that too many houses existed. Could anything be more absurd? Surely nothing except the proposed remedy, namely, to issue yet more of the very kind of base money that had been, all through this period, the most malignant cause of poverty, depression, and loss.

Since the passage of the legal-tender acts in 1862 and 1863, the so-called money of the United States in common use has been bad money. It is still bad, though in lesser degree; and it will continue to be bad and to work subtle mischief, until coin only shall be lawful money and legal tender for debts incurred.

In order to prove these dogmatic propositions, and to make the use of money and the function of banks and banking perfectly clear, we must analyze the simplest transactions, then proceed from the simple to the complex; and last we shall see, if we succeed in the analysis, that the metaphysical instruments of exchange, which are known as bank-notes, bank-deposits, bank-credits, and bank-exchanges or clearances, are as essential to the quick and cheap distribution of corn, beef, pork, and cotton, as the railroad, the steamship, the butcher's wagon, or the baker's cart. It may, I trust, become very plain to you that, unless these instruments of exchange are convertible into the coin which they represent, their service is impaired or lost.

If we analyze the simplest exchange, we find that all transactions are of the nature of barter. To go back to school-boy language, all trade, from the transaction in the proverbial jack-knife to Vanderbilt's great sale of twenty-five million dollars' worth of railway-stock, is nothing but "swapping." Why do we swap? In order to get more than we give, *i. e.*, something of more use to us than what we give; here begin the metaphysics. The exchange occurs because there is a mental conception that the things bought will be of more service to the buyer than the thing sold; hence the conception of value. Each person buys and sells. The man who sells corn buys money; the man who buys cloth sells money. The equation may be formulated in words as "service for service," in which the conception of price arises as the mean of the equation. The dollar is the common factor.

When the mental conception of service is applied to substance, then the equation takes the form of "product for product." Carry the mental conception a little further and we at once perceive that, in order that any exchange shall happen, another formula must be conceived, and that is "effort for effort."

We may use these words rather than "labor for labor," because the word "labor" has become limited to muscular or bodily work upon material substances, while effort includes that, and also, in addition, the mental functions or efforts that are serviceable to others, and for which something will be given in exchange. No one but a fool sells something for nothing. The mistake which the labor reformers make is in not admitting mental effort as one of the highest forms of service.

The process which must occur in order that any exchange, barter, swap, or other dealing between men shall happen, must be a purely mental consideration of the effort exerted in the production of the thing parted with, and the effort saved by becoming possessed of the thing obtained. It may be unconscious cerebration; but even in the proverbial knife-trade, each boy swaps his knife because he thinks he gets a better knife than he gives. In the boy's case there is usually a misconception on one side or the other; but in the great commerce or swapping of the world, whether among men or between nations, each does obtain that which is more serviceable than that which is parted with, or else the traffic ceases.

In the last analysis all commerce is an exchange of reproductive forces.

All consumption is a conversion of forces. In the end it is a chemical reaction; and, the wider the distribution, the more perfect the conversion. All this is elementary, but yet necessary to the further treatment of the subject.

Exchange is necessary to the subsistence of the human race, and some kind of money is necessary to facilitate exchange.

The point most commonly overlooked in commerce is that two and two make five,—sometimes six, and even more,—and the units over are divided sometimes in equal portions, sometimes unequal, between the parties to the transaction. The force of the grain stacked upon the wheat-field, and of the cotton on the plantation, are both passive. Convert them in the factory, and an active force is developed which serves to clothe the bodies of men. Two measures of wheat and two measures of cotton make five measures of cloth. The cotton on the field is useless to the producer; the wheat may rot upon the prairie. Bring them together, add the work of the factory, and by their conversion the new force is developed that is measured by a higher price in money than the prices of all the elements of which this new force consists. You will observe that money does not constitute one of these forces, or one of the elements of the new force. It is only an instrument used in their conversion.

Let us now assume that the nation has had the intelligence to adopt the best kind of money yet discovered, namely, coined gold, as its standard of value and only legal tender, and coined gold and silver as its instruments of exchange or its money.

Into the somewhat abstruse question of the bi-metallic theory, and the ratio of gold and silver to each other, I do not propose to enter. Let us assume that the legal-tender acts whereby United States notes have been made lawful money and legal tender, have been repealed by Congress or annulled by the Supreme Court. We then stand ready to begin the consideration of the subject of banks and their relation to the railroad as the agents of exchange.

In order to be sure of our ground, we must begin *ab initio*. For this purpose, we will consider the traffic in black pepper. Pepper is produced in the island of Sumatra. Down to a comparatively recent period, the natives of the island had not developed wants in respect to the products of civilized countries to a sufficient extent to balance the traffic in pepper without the inclusion of a considerable amount of money—I mean, of course, real money—in the transaction. And here you will observe that in all international trade there must be an exact balance. No nation can sell unless it buys, or buy unless it sells: and what is called the balance of trade is and must be only the balance of gold or silver coin that is bought or sold. These coins are commodities, products of labor, of precisely the same kind as beef, pork, wheat, corn, and cotton, and subject to the same laws. In the year 1879 we bought of foreign nations about eighty-four million dollars' worth of gold in the form of coins; that is to say, we bought English, French, and American coins made of gold, weighing a certain number of ounces, of which weight the stamps on the coins were the certificates. A true statement of our foreign traffic, taking for the moment no consideration of credit or payment deferred on either side, would be that,

We sold *so many* bales of cotton,
——— bushels of wheat,
——— gallons of oil,
——— pounds of meat.
We bought *so many* yards of cloth,
——— tons of sugar,
——— bales of hemp,
——— ounces of gold.

Assuming all transactions to be on a cash basis, there can be no balance of trade. The exchange is an exchange of equivalents, but each party assumes, and, on the whole, does make, a profit; that is, each nation parts with that which it could not use with as much advantage to itself as that which it receives. There is an exchange of forces, but in this exchange two and two make five, and the one over is shared by the two parties. Sometimes one nation makes a larger profit than the other; but both must gain something, or else the trade will stop.

Let me call your attention to one point. After a nation has coin enough for bank-reserves and for use as pocket-money, the most unprofitable thing it can import is more coin. The only use you can make of the excess of coin is to send it out of the country again. You cannot consume it. All other goods and wares you can convert into some other useful form, but gold can only be made into jewelry, and silver into table-ware.

When there is no balance of trade, so called, that is, when our cotton, grain, and oil are equivalent to dry goods, sugar, and spice, then the conditions are very sound and healthy. If, on the other hand, we sell what is worth a million to us, and in exchange appear to get what is entered at the custom-house at a million and a quarter, then we may be borrowing the excess. Or if we export more in value than we import, we are either paying our debts or losing by the traffic, and yet this last state of the account is commonly called a trade that "shows a favorable balance." The coin we imported last year we needed, but this year we need iron, salt, sugar, etc., and the import of coin has diminished. The so-called favorable balance has diminished; our demand for these things is giving our best customers for our grain and meat more ability to buy; they can pay in iron when they could not pay in gold; but I do not hear any complaint of adversity because the balance of trade has changed. We bought gold when we needed it, and paid with cotton, wheat, and. oil; now we want iron, wool, and tin, and we are buying them in the same way.

Let us return to the pepper. The natives of Sumatra could not use all their pepper; there was an over-production of pepper there; they 'had very little use for American goods, but 'they could use good money. These people, however, wanted a particular kind of money. They had learned in some rude way that, whatever faults the Spanish nation had committed, their coined dollars, known as " Carolus " or " Pillar" dollars, always contained the same amount of silver: therefore these dollars they would take; they would swap pepper only for Pillar dollars. And hence it happened that the American merchant could only get pepper by sending his ship partly loaded with goods, and the rest in ballast, with Pillar dollars for the balance in order to buy pepper. How the pepper traffic is now carried on, I do not know; this was the way when I was a boy. This is still the rule in respect to a large part of our traffic with China. For a very long period we settled our balance of trade, so called, with China in Mexican dollars. That is, we bought silver in Mexico, and sold it in China by the measure of the dollar. Here is another curious anomaly: Mexico stands as the example of all misgovernment, anarchy, and confusion, but Mexico never debased her coin. Is there not hope for her? We, however, at length obtained the confidence of a small portion of the Chinese, so that they were willing to take our trade dollars, and now we sell China a good deal of silver in that form.

You will observe the most costly method of commerce in these two examples: special kinds of coined money to be gathered up, packed, and shipped across the seas, subject to all dangers of loss by Malay pirates but a little while ago, and to all the constant dangers of storm and shipwreck for all time; the ship perhaps making a voyage half around the world almost empty in order to bring home the pepper or the tea. You can readily see how limited such commerce must be.

Transfer these conditions to our own land; suppose that the Louisiana purchase had never been made, and that Iowa, Nebraska, Kansas, and all that vast Mississippi valley belonged to a foreign nation, and were separated from New England by a line of custom-houses more costly and difficult to pass than the Hoosac Mountain or the ridges of the Alleghanies; assume that there was no mutual confidence, and that each nation watched the other with jealousy and suspicion from behind ramparts guarded by five hundred thousand armed men, with a yet greater number in reserve, wasting even in the reserve as much time in drill with rifle and sabre as they would spend in work at the plough, the loom, and the anvil. When you have assumed these conditions, you have only made a comparison with what are called the civilized nations of Europe, omitting Russia and Turkey; who, with only four times our population, now stand thus facing each other with two million men in camp and barracks, a larger number in reserve, bound in the fetters of sixteen billion dollars of national debts secured by mortgage upon a territory only one-half as large as ours, omitting Alaska. Study the history of these countries, and you will find that, in former times, commerce could only be carried on between them by the actual movement of the coin; and that most of their wars have been incurred, and their great debts imposed, because the beneficent function of commerce was denied, and because each tried to gain a special advantage over the other without rendering a service in return.

But you will say: These obstacles to mutual service do not exist on this territory; what have they to do with banking? I only refer to them to make the contrast greater. Even these contests of race and differences of institutions and language would not restrict the exchange of corn for cotton, of beef for iron, of wheat for fabrics of every kind; would not be as great obstacles to commerce as those that are removed by the existence and use of banks, and by the service of bills of exchange, bank-notes and checks, and bank clearing-houses.

Suppose you could not get a barrel of flour from the West without sending out a five or ten-dollar gold coin to pay for it, and then you begin to see the function and use of banks and bankers.

By the use of a little slip of paper inscribed with a few words, signed by a responsible bank-officer, the title to one barrel of flour passes from the farmer in Nebraska to the mechanic in Massachusetts; while the title passes from the mechanic in Massachusetts to the farmer in Nebraska, to a certain number of grains of gold minted into coined money in the works of the government at Philadelphia.

The coin may be all the time kept for safety in the vault of the Sub-Treasury in New York, and the barrel of flour may be stored in a warehouse in Chicago for months before it is consumed; but the title is passed from one to the other by the assignment of the little strip of paper, inscribed, " The Merchants' National Bank promises to pay to the bearer five dollars on demand," signed by the president and cashier; or by another strip of paper in similar form :—

MERCHANTS' NATIONAL BANK.
  Pay to the Iowa farmer five dollars on demand.
    TO THE CASHIER.   (Signed)   JOHN SMITH.

This is an epitome of all transactions. The bank is the agent for assigning and transferring titles to property: that is the exact function of the bank or banker, nothing more and nothing less. The property assigned may either be its own capital in coin, or a title to some property of its depositors. A part of its capital is kept in reserve in the form of coin, in order that if any one wants actual money,—true money, coined money,—it may always have enough to meet that demand. It lends the rest of its own capital, and it acts as agent to transfer titles to the capital of others.

If these functions are carefully considered, it will be observed that the abundance of notes, checks, bank-deposits, bills of exchange, and other instruments of credit by which titles to actual property are passed from one to another, will be in exact ratio to the quantity of capital, that is, of commodities or property, thus being moved or assigned at any given time. This property, these commodities, constitute what is called the quick or active capital of the community, consisting of beef, pork, hay, corn, cotton, dry goods, tin-ware, boots. Bear in mind, we are not now considering savings institutions, also called banks, that deal more in titles to fixed capital, buildings, works, and improved lands, but we are considering the functions of commercial banks and bankers who serve the purposes of merchants and manufacturers.

The interest which is paid, and which constitutes the profit of the bank or banker, is paid in a limited degree only for the use of money: no actual money has passed; the money is substantially all in the bank-vault, or in the vault of the Sub-Treasury; the interest is paid for the use of the property, of which the bank-note or credit has passed the title from the lender to the borrower by the measure of money. This property is a product of labor; interest is therefore paid for the service of labor already done in the past, in order to enable the borrower to perform more work in the present.

When you mortgage a house to a savings bank, what do you borrow? is it not a part of your house? You are a mechanic, and have saved five hundred days' work for which you have a thousand dollars in gold coin; you spend that, but you want more house; you borrow a title to another thousand dollars, and buy with it five hundred days' work of other mechanics, to finish your house; and you owe the sum that you have spent until you can work it out yourself, but you have really borrowed half your house.

Prosperity consists in the rapid consumption of the goods and wares that I have named, meat, flour, pork, iron, cotton, and the like.

When the money in use is good money, such as gold coin, that only changes its value in relation to other products of labor in long generations, then confidence will be sufficient to promote the quick circulation of commodities, and then will follow the consequence so often mistaken for the cause, —there will be a great abundance of bank-bills, bank-deposits, and bank-credits; every one will say, "Money is very plenty;" but the real fact may be that the amount of real money held in reserve to meet emergencies, in the vaults of the banks, and the Sub-Treasury, may not have changed a single dollar; but, the money being good, productive and constructive enterprise will be active because confidence is assured.

In such periods it is capital that is plenty,—iron, beef, cotton, potatoes, pepper and salt, milk, butter and cheese (the annual value of our dairy product is greater than that of our cotton), and we work cotton into cloth in order to obtain butter, cheese, eggs. *De minimis curat economicus.* When capital is abundant and confidence is great, the new railroad is projected, the new mill is constructed, the new house is planned, and we spend or consume the products of the year present, in order to be able to provide for the wants of the years to come. We convert the perishable forces of the year present, that would otherwise decay, into the more permanent forces,—into railroads, mills, and works that will assure more abundant production in future years.

When your money is not true, that is, when it is subject to the caprice of Congress, people live from hand to mouth, and the work that is necessary to be done by each generation to prepare for the increase of the next is stopped, because the money that may be received in the future may not measure the effort of the present. For several years after the panic of 1873 we lived as if never another mile of railroad, or another factory, or another house, would be wanted; the portion of the population usually employed in providing for future need was reduced to idleness,—may be five in a hundred; wages were depressed, the stock of goods piled up, and wiseacres talked of over-production; then in the next breath they would say we must save, and not spend. Why! the very thing needed was that we should spend our excess of iron and copper, of corn and pork;

spend them in new work. That is just what we are doing now. Money was said to be plenty in State Street, and would not bring three per cent. per annum. But what was this money? It was the title to these unspent commodities that no one had confidence enough in the future to use or spend, because the measure of spending, the money, was bad.

On the 1st of January, 1879, men came to believe that the standard of value had become fixed, that specie payment was resumed, that gold coin had become once more the money of the nation. Confidence returned, and now what do we see? We are spending again in useful work; we are converting iron into railroads and machinery; brick and timber into mills and works. At the same time our stock of real money, held in reserve in gold coin, has increased more than one hundred million dollars.

As soon as we began to use good money, it flowed in upon us. We have ceased to hear of over-production, yet the products of 1879 were the most abundant ever known.

It is our mental condition only that has changed. Now that good money is even partially assured, we find our force is doubled; industry is resumed, and labor is well employed, because confidence is restored.

In order that we may more readily comprehend how these little strips of paper that I have described—these checks and bank-notes—really do their work, let me use a word very familiar to those who, like myself, have been book-keepers, the word "cash." If you ask me now, "Have you any money in your pocket?" and I followed my own rule, I should confine my answer to the coin in my vest-pocket; but if you asked me if I had any "cash," I should also include the bank-notes in my pocket-book.

In book-keeper's parlance, "cash" consists of checks,

bank-notes, United States notes, and coin. A book-keeper never says his money is short, when he cannot square his account ; it is always, " My cash is short."

I suppose none of you know what it is to be short of cash ; if you do, you are probably not very particular what word designates the instrument by which the deficiency is covered.

The cashier used to be the guardian or keeper of the " caisse," or chest ; he was the chest-keeper, in which coined money was kept by each merchant when banks were few or none. Now his chest has disappeared, and he keeps a cash-book, in which titles to money are registered ; and, in place of coin, he balances his account by means of the notes and checks by which the titles to money or to other property measured in money are passed from man to man.

How do we use this "cash" as a substitute for money?

The other day I wanted some smoked venison-hams, such as are brought into St. Paul, Minnesota, from Pembina, where deer and Indians abound. I knew no one in St. Paul who would sell me hams unless he had "cash" in hand. What did I send him? Not a piece of gold ; that would have been foolish, although I had three ten-dollar pieces of gold in my pocket, that I had drawn from a banker, in recompense for a lecture given to this club last winter and afterward published in an English review ; that is to say, I had some true money—some capital in gold coin.

I took that money to a bank, and obtained a cashier's check on a bank in New York. I parted with my three coins, and obtained a title to, or draft for, other three coins of same denomination; that is, containing the same exact weight of gold. I sent that title to the provision-dealer in St. Paul, and by the next train of cars came back the smoked vension-hams, cured by the Indians of Pembina.

Money might have been said to be plenty in St. Paul, to the exact amount of the three ten-dollar gold coins; but the coins themselves were in the vault of the bank in Boston, to whose cashier I paid them for the draft. The Indian had brought in the hams to the shopkeeper in St. Paul, and had exchanged them for blankets, gunpowder, bullets, and probably some whiskey, for which the shopkeeper owed the manufacturers of whom he had bought his stock. In this transaction you have an epitome of all commerce: the shopkeeper in St. Paul received the title from me to three gold coins,—not the money itself,—and sent me hams; he swaps ham for a title to gold; he deposits that title to gold in his own bank in St. Paul, with other "cash" received for goods; then he draws his own check on that bank, and pays his own debt for blankets and gunpowder: and so the title passes from hand to hand, and from bank to bank, until, in the clearing-house of New York, one check is balanced against another, and a little specie or real money passes from one to another to settle the balance.

My small mental effort procured the gold for me, and the Indian's gun procured the ham for him. In the consumption of the ham the substance of my brain was restored, after the effort which found its expression in the English review, so as to enable me to make this effort to explain the science of banking to you; while, in the consumption of the whiskey, the Indian obtained a gratification, and, in the use of the blanket and gunpowder, he was fitted out for another hunting expedition.

The circulation of the commodities called the bank-check into existence. "Cash" was plenty in St. Paul to the extent of that check; it served its purpose in liquidating other transactions; but the only "money" transactions in the whole sequence was the movement of three gold coins from the vault of Kidder, Peabody & Co., in State Street, to the vault of the Eliot National Bank, in Devonshire Street.

How did the coin get into the vault of Kidder, Peabody & Co.? Perhaps as a part of the $84,000,000 sent here from England in 1879, in exchange for Minnesota flour, ground in the mills of the same city of St. Paul; or perhaps it had come as a product of the labor of the miner in California, which he had parted with, in order that he might purchase the cowhide boots of East Brookfield, or the heavy woollen blankets made in some Massachusetts factory.

The elements of banking might be put in a formula, almost in a scale. They consist of:—

A little gold coin or true money.

An unmeasured amount of character, prudence, forethought, and integrity, in the banker.

An unlimited amount of confidence on the part of the community.

The scale cannot be given in adequate terms. For this country, it might now be stated something like this:—

Three hundred million dollars of gold coin suffices as the standard by which to measure three hundred thousand million dollars' worth of purchases and sales every year.

By the use of notes issued by, or checks upon, banks and bankers, more than 100,000,000 tons of food are moved in each year from the producer to the consumer, and thus the subsistence of 50,000,000 people is assured.

This is the power of true money; this is the moneypower. This is the work that knaves and sentimentalists denounce, obstruct, and retard. This is the measure of the integrity of men; the measure of the trust that each man reposes in his neighbor; the standing testimony that total depravity is but the gloomy dogma of the shallow thinker, whose insight into the great work of the world is but the depth of his own little mind.

The great crops of this country—grain and hay only—weigh 100,000,000 tons; they constitute food for man and beast,—two tons to be moved from field and pasture to subsist each man, woman, and child; moved not once, but twice and thrice. The grain must be moved from field to railway, from railway to mill, from mill to warehouse, from warehouse to baker's oven. The hay and roots must be moved from field to stable, be turned into butter, cheese, and meat, be exchanged for sugar, tea, coffee, and spices; each kind must be distributed, worked over, converted from one form into another, and at last consumed. The mind cannot conceive the exchanges that take place each and every day.

The money lies safe in the vaults of the great cities, but the little slips of paper, by which a title to it is passed from hand to hand, serve all the purpose, provided only that the money is good, and that bank officers are honest and prudent men. There is no better measure of the character of a nation than the use it makes of banks.

We can only approximate the work that must be done in order that each of you may subsist a single year. Two tons of grain and hay to each one, partly used directly and partly converted into meat: each of you eats more meat than flour; then come the milk, the sugar, the vegetables, the coal to cook the food and warm the house. All this conversion of force must take place that you may not starve,—not less than three tons weight, six thousand pounds, moved at least three times; first, thousands of miles, then hundreds, and at last, half a mile as to each small parcel. This work must be done every year for every one of you,—too much work done for the value of a freshman, some of you sophomores may think.

All this dead weight must be moved and recombined, that each of you may subsist; and if the work stopped a single year, or even half a year, the world would be depopulated. A snow-storm in London reduces hundreds to the verge of starvation. And through all these changes the little strip of redeemable stamped paper, with a promise to pay upon it, and signed by one or two names,—the bookkeeper's "cash",—has been a sufficient instrument to serve all this vast and complicated traffic; the bank-note, the bill of exchange, the bank-deposit certified by a few figures in a book, with a little coin to make change and settle balances, has measured each change of ownership, and has passed the title of all this property from man to man; while the railway, the steamship, the butcher's cart, and the grocer's wagon, have moved the property itself. There is not coin enough in the world to do this work alone; but without the coin to serve as the standard by which to measure and guage all this traffic, it would mainly cease. The whole mass of gold in the world, the painful accumulation of centuries, valued and sought by every race and every nation since the dawn of history, would not fill this hall. The one product of labor that neither moth nor rust can corrupt, that neither air nor water will oxidize,—who can tell when its service first began, or how it came to be used as money or the standard of value?

Can you find a deeper problem in metaphysics than the analysis of the conception of value,—the estimation of gold,—the twofold process of the mind which seems so simple when we buy and sell, but is so subtle? If you can follow the course of the little slip of paper stamped with a promise to pay dollars, as it passes from hand to hand, and carries with it the title to the hundred million tons of food, until each daily ration reaches the mouth that is to consume it ; if you perceive that as each ton moves by rail and river, the paper slip, the book-keeper's "cash", passes by mail and hand ; if you can see that the volume of little slips and the sum of the figures on the ledgers of the merchants and the banks, mark as many dollars of promises and credits as there are dollars' worth of merchandise moving from producer to consumer,—then you will have mastered the first lesson in banking ; and I may tell you perhaps, privately, that you will know more about it than ninety-nine bank-directors in every hundred.

If you will try the experiment, you will find that nearly every practical man will tell you that banks borrow and lend money, and will be amazed at your audacity if you deny it; but at the same time they will admit that neither a bank-note nor a bank-check nor a bank-deposit is money.

Does not this speak well for the general integrity of men, that more than ninety-five per cent of all the transactions of life—the exchange of the hundred million tons of food that I have named ; the conversion of this force into the thousand forms that make up the necessities, the comforts, and the luxuries of life ; the whole traffic on which the subsistence of nations depends—are worked by means of little slips of paper that merely carry directions from one book-keeper to another how to make up the merchants' and the bankers' accounts, so as to show by the trial balances who is in possession of the property exchanged, or who is consuming it at any given time ? You will observe that these transactions are world-wide. The bill of exchange that passes from nation to nation is but another slip of paper by means of which a title is passed. Even yet more wonderful is the telegraph. It almost passes comprehension when we witness its work. The tea merchant in London sends one message to China ordering tea, and another to San Francisco for silver, and before the week is ended both substances are on their way from the producer to the consumer. Two clerks make their entries, two letters of advice are written, and in the London banker's office the transaction is settled.

It is important to impress upon your minds that banks and bankers transfer titles to consumable commodities from producer to consumer; and, further, that in the consumption of the commodity by the consumer is developed the force to produce some other thing with which the first producer is paid. The title passes by a written or printed slip that is but the certificate of "cash" in the book-keeper's accounts. Nearly all the so-called money that passes is a direction from one clerk to another how to make an entry on his ledger. I have repeated this formula many times, and have tried to make it plain; it is the essential idea that must be comprehended.

It follows of necessity, if the system of banking is sound and bankers are prudent, the sum of the bank-notes, bank-deposits, and other forms by which titles are transferred to property on its way to consumers, can never exceed the nominal value of the commodities: hence money is said to be plenty or otherwise, when the quantity of commodities is abundant or otherwise. The danger to banks and bankers comes when prices have been carried to a very high point, and begin to decline slowly or quickly: then comes the doubt whether the men who have borrowed titles to cotton or wool or other merchandise, through the intervention of the banks, can convert these materials into cloth or the like, and obtain by its sale a title to as much as they have expended.

The doubt begins with cautious men, spreads slowly or quickly; if the activity has been very great, if the substance borrowed has been wasted in useless mines, or spent in constructing railways that are not yet wanted, then panic may ensue; each depositor fears his title will be passed to some one who will not use it wisely; then a run is made upon the bank to convert the deposits into money, and withdraw gold from the bank. These crises come usually for good reason; they are the process of cure, not the disease itself: the disease has been the wasteful or injudicious expenditure of the substance long before borrowed; it has been the imprudent lending of titles to commodities to those who in consuming the commodities have not reproduced something that is salable; who have spent them without results.

Let us now consider the work of a national bank.

The process of organizing and working a bank is very easily comprehended when the fundamental idea is grasped, that a bank lends its own capital, and transfers titles to the capital or property of its depositors.

A portion of its capital it must always keep in its vaults in coin, as a reserve. How much that reserve should be, depends upon the kind of business done by the bank; and the proportion of reserve is an indication of the prudence and skill of the manager.

Let us assume that the capital of a bank has been paid in by its stockholders in gold coin, say **$1,000,000**
The bank proposes to become a national bank, and it at once lends one-half of its coin to the government at four per cent, interest, for which it receives bonds 500,000
It has left in coin 500,000
On the deposit of the bonds as collateral security for the notes it may issue, the government then authorizes it to issue national-**bank** notes for the sum of 450,000

What is a national-bank note? It is a promise of the bank to pay to the holder a certain number of coined dollars on demand. The notes of the bank, when in its own possession, are therefore unused evidence of its own debt, and are of no effect until issued. How do they get into circulation?

A manufacturer who has made ten thousand dollars' worth of cloth, and who has not paid for the wool or the labor, desires these notes to use for the purpose of such payments. You will observe that they are promises to pay coin, and the bank has in reserve half a million of coin. These notes are therefore transferable titles to a part of that coin.

The manufacturer has sold the ten thousand dollars worth of cloth, for which he has not yet paid, to a jobber, for eleven thousand dollars, and has taken his note at four months for it. The jobber has the cloth ready to sell to the consumers: the consumers are in part wool-growers and mill-operatives. The note is a title to the equivalent of the cloth in coin; the sale of the cloth will enable the jobber to pay the note. Therefore the note of the jobber is a title or evidence of the existence of so much cloth on its way from the producer to the consumer.

The manufacturer takes the note, due in four months, to the bank, to be discounted; the president deducts interest at whatever the market rate may be, say at six per cent. or two per cent. for four months, and gives the customer $10,780, in its own bills or promises to pay on demand. In that discount of interest is the profit to the bank; the manufacturer pays for the wool and the labor $10,000, and has $780 left in bills. He now wants some foreign wool, for which he must pay gold. He presents $780, bank-bills, and draws that amount from the bank's reserve of coin; the rest of the notes circulate from hand to hand; some of the farmers and operatives who received them from the manufacturer buy goods of the same dealer who purchased the cloth; by the time his note is due he has received these bills, and has deposited them in the same bank that owns his note, and, when the note is due, draws his check, and thus pays, or offsets his deposit-account against his note.

While the note has been in existence, the cloth has been in use; it has enabled those who wore it to do more work, to reproduce other capital to take its place.

All through the transaction the gold has been in the bank, ready to redeem the bank-note; the cloth has been reproducing capital, to assure the payment of the merchant's note. The bank-note and the merchant's note have divided the title to the gold and the cloth, and passed it to a hundred different hands; but the issue and redemption have been worked to the convenience and profit of each and all.

Confidence and credit and a few slips of paper have removed the need of weighing out gold for wool, and wool for cloth, and cloth for labor. The title has been passed, and all the work has been done, because men can trust each other; the slips of paper have carried the title, and enabled the book-keepers of the banks and merchants to keep their record of credits granted and obtained; and, in the clearing-house, one slip written off against another squares the account. Coined money has been the standard; convertible paper money has been the instrument; an entry in a ledger has been the conclusion.

In order that the conclusion may be just and true, the substance to which the title has been passed must have been rightly spent; more force must have been generated than has been consumed. The difference will have taken the concrete form of a new and useful railway or mill, a better house, a college gymnasium, or a Boylston Hall, in which students may be making preparation for more effective work in the future. Thus the world goes on, never more than one year removed from starvation, yet with always enough and to spare. Whether that which would suffice shall be where it is wanted, or not, is no longer a question of physical means: railroads and steamships can assure distribution to almost every part of the world. The conditions of prosperity are now peace, order, and good-will among nations, good money, honest and prudent bankers. When the interdependence of nations is admitted, then, and only then, will commerce forbid war.

I have stated to you that our great crops of grain and hay weigh more than one hundred million tons. The hay is only a partial measure of the meat, the butter, and the cheese; the roots add yet more. One hundred and fifty million tons of food is within the measure of what we consume ourselves, or send abroad to exchange for goods and wares of every sort,—three tons to each man, woman, and child, to be converted into power. Food is fuel for the human engine. "Going into business," which some of you may contemplate, means a share in the conversion or distribution of this force of three hundred thousand million food-pounds.

What was your share to-day? About sixteen and a-half pounds: three consumed directly, the rest indirectly. Witness the power of money: that it must be an accurate measure of the division of three hundred thousand million food-pounds into daily rations of three pounds each. Legislators in Washington are now tampering with the standard of value, and attempting again to alter the measure by which all this vast traffic is to be conducted.

You may see how little we are governed,—how much we may be misgoverned,—when you attempt to conceive of the mischief that would be done if all the rules by which this work is accomplished needed to be established by statute. Do you not see that when any attempt is made to extend the function of statutes beyond the enforcement of justice and the collection of the necessary revenues, with right provision for education, it must almost of necessity raise barriers between men and nations that would have no existence in the nature of things? Honest men need no statutes for the conduct of their business: the statute intervenes only when some one tries to get an advantage over another; in other words, tries to obtain more service than he renders.

One by one all sumptuary laws have been repealed, or have fallen into disuse, because trade makes its own laws. If a tariff for taxation is assessed at rates beyond a certain point, the smuggler renders it inoperative. Attempt to collect two dollars a gallon on whiskey again, and the revenue on it would almost cease.

Issue fiat money, and who would exert himself to become possessed of it? Only the man who believed he could cheat his neighbor by inducing him to give something for it, or who would force him to take it, under the operation of a legal-tender act, in place of the true dollars that he had promised. Show me an advocate of "fiat money," and, in nine cases out of ten, I will show you a man who either desires to cheat his creditors, to grow rich by causing other men to become poor, or to live without work on the product of some other man's labor.

I shall now be obliged to lay aside my strict definition of "money," and the limitation of that word to coin, and fall into the customary way of treating convertible bank-notes and legal-tender notes as money, or, in common speech, as "paper money;" better designations are, in respect to coin, "real money," and in respect to convertible paper, "representative money."

Notes serve the purpose often given as descriptive of money; they are instruments of exchange; and it would be almost a Quixotic attempt to strive now to change their common designation. We will call both classes of notes, "money," in order that 1 may more fully explain why one is good paper money and the other bad paper money. Both are promises of coined dollars on demand, but the redeemable bank-note is the symbol or measure of the cloth, meat, corn, cotton, or some other substance, on its way from producer to consumer. It can only get into circulation, as I have attempted to explain, as a representative title, or evidence of substance, in the consumption of which will be given the power to redeem the note.

The legal-tender United States note, on the other hand, is the symbol or evidence that the government forced its citizens to lend it food and munitions of war fifteen to twenty years since, all of which were consumed without reproduction; it is evidence of capital destroyed, and of debt due and unpaid. Its convertibility into coin depends on the power of taxation. It has not the first attribute of good paper money, except so far as coin is held in reserve for its payment; nor has the government any immediate means of payment, if any sudden distrust should cause the notes to be presented beyond the sum of its reserve in coin. In banking, the proportion of reserve can be determined by the nature of the business done, the condition of the crops, the state of the foreign exchange, and many other indications, a knowledge of which constitutes the skill of the banker; but the safe measure of reserve for a government note can never be less than dollar for dollar in coin, and, when that standard is established, the issue of the notes yields no profit or saving of interest.

In conclusion, let me indicate one other advantage which a national-bank note possesses over the notes of the State banks, formerly used. The State-bank notes depended entirely on the skill and judgment of the bank managers: when a bank failed, the holders of the bank-notes had a lesson in the meaning of words; they found out to their cost that notes might cease to be *money*, either in fact or in semblance.

State banks often failed to pay their notes as well as their deposits.

The national-bank note, or promise of the bank, cannot be issued unless the bank has first lent a part of its capital to the government, for which the government pays interest, and in evidence of which it has issued bonds. These bonds are deposited as security for the payment of the notes. The bank may fail, it may defraud all its depositors of every dollar of the title to capital which they have deposited with it, but it cannot defraud the holder of a note; if the bank does not redeem the note at its own counter, the holder can present it to the controller of the banks, cause the bonds deposited as security to be sold for coin, and draw the coin. The bank-note is secured first by all the other capital and profits of the bank not lent to the government, by all the commodities in title to which it was first issued by the bank and obtained circulation in the community, and, second, by the collateral security of United States bonds bearing interest.

The United States note depends upon the power of future taxation, and is at the caprice of Congress, into which such men as B. F. Butler have more than once found an entrance by the votes of their dupes and their confederates in Massachusetts and elsewhere. It does not represent property in existence. but substance that has been destroyed.

Which of these notes best meets the conditions of safety?

May it not be affirmed that the national-bank note leaves nothing to be desired, if paper money convertible into coin is to be used at all? It is secured beyond a reasonable doubt, and as it has the semblance of true money to masses of people who cannot appreciate the distinction between real money and its promise, it is eminently right that the government should protect the holders of the notes, and assure their absolute convertibility on demand by requiring the deposit of the United States bonds as collateral security for the notes.

We have, indeed, brought United States notes to par in gold coin, and for the moment he who presents them for payment will receive the coin; but if the preceding statement of the function of banks and of bank-notes has any foundation in principle, the attempt of a government to assume the functions of a bank of issue is an economic absurdity fraught with the gravest dangers.

The question is not yet determined, but is still at issue, whether the money of the nation shall be good or bad for the next few years.

The lawful money is now *good* money in gold coin, and *bad* money, or United States notes first issued for the purpose of collecting a forced loan, and made a legal tender for that purpose only.

During the war these notes depreciated to less than forty per cent. of their nominal value; they are now at par, and are nominally redeemed in coin; but although the lawfulness of their reissue is contested by the ablest lawyers and the members of the Senate and House of Representatives most competent to decide the question, they are being reissued even while the validity of the acts under which the reissue takes effect is before the Supreme Court for adjudication, it being a question not yet decided. Their reissue is not confined to the purposes for which the executive might feel obliged to use them under existing laws, but they are being forced into use again in the purchase of bonds not yet due, for the sinking fund, without reason or necessity.

This course is but a repetition of the disastrous policy followed under the administration of the Treasury Department by most of the predecessors of the present secretary ever since the office was held by Hugh McCulloch. When these notes which have been paid in coin are reissued in exchange for bonds, such notes being legal tender until otherwise decided by the Supreme Court, and therefore competent under existing laws to constitute a portion of the bank reserves in place of coin,—they, in fact, constitute an element of the currency not called into use by the operation of the laws of trade.

They are therefore forced into use where they are not required, and may at any time work the same effect that they did before, to wit: inflate prices, and presently cause the export of the gold coin which will be displaced by them. Next may follow their depreciation, and possibly another suspension of coin redemption by the treasury of the United States; or what would be a yet greater misfortune, redemption in depreciated silver coin.

The first steps in this vicious sequence are now apparent, and the malignant effects of the attempt of the Treasury Department to do the work of a bank of issue, for which it is radically unfit, are now to be as plainly seen as they have been many times before.

Speculation waits upon the decision of the Secretary of the Treasury as to how much bad money he will inject into the currency in each week; and the eaves-droppers of the lobby listen for the corrupt whispers that shall enable them or their confederates to plunder the victims of a false monetary system

The prices of the necessaries of life have been subject to great fluctuations, as they have before when the currency was tampered with. In 1879, they rose faster than the wages of those who did the work of producing them, and strikes prevailed everywhere; the unwary were again misled by the specious representations of those who live upon the credulity of their dupes, and the thousand evils of tampering with the money of the country became patent to those who look beneath the surface. Mining stocks were sold at such prices that if the product of the mines would pay a dividend on the nominal sums given, silver would be depreciated at least one-half from its present ratio to gold; any thing that was called a railroad served the purpose of the stock-jobber, and many of the other symptoms became visible which constitute the disease of which a commercial crisis is the usual process of cure.

These are the symptoms of a false element in the finances of the country; of bad money again displacing that which is good.

Whether an inflation caused by the use of government legal-tender notes nominally redeemable in specie, and not cancelled when thus redeemed or paid, but reissued, will work as great a disaster as the inflation caused by the forced circulation of the same notes when irredeemable, is one of the problems not yet determined.

The enormous crops of the past few years, and the possibility of moving them which the railroad and the steamship have given us, have enabled the Treasury Department to meet the conditions of the resumption act, and to stand ready thus far to redeem the notes in gold coin when presented. A true statesman would be able even now, to assure the stability of coin payments for all time to come; but, to the shame of our intelligence as a people, it is yet a question whether another financial disaster may not be needed, before the simple principle of finance is learned, to pay your debt due on demand first and finally, rather than to reissue your own evidences of debt due on demand, and force them into circulation as lawful money in the purchase of long bonds not matured.

If we are saved from another disaster which may come because of the want of capacity on the part of those who assume to govern and control the finances of the country to comprehend, or their unwillingness to accept, the simple principles that underlie the question, it will be from the same causes that have brought us into our present favorable condition in spite of previous mismanagement.

The enormous productive capacity of the country and the energies of the people, aided by the railway system, have enabled us to surmount financial incapacity, under previous administrations, equalled only by that charged on the Tory administration of Great Britain by the great leaders of the Liberal party.

Full credit may be given to the present Secretary of the Treasury for executive ability and administrative power. The conduct of affairs has been admirable during the period when the circumstances of the time—our great harvest, and the bad crops in Europe—gave us, for the time, the control of the gold of the world.

But the point of danger is near or is already reached; the test of statesmanship is now being applied. Circumstances may again save us, but the reissue of notes already paid, after the disastrous experience of years past, caused by the same vicious policy, may fully warrant those who resisted that policy then, and foretold its malignant result, in again sounding a note of warning.

The danger of a debt currency must exist so long as the promise of coin is *forced* into use by an act of legal tender. Such a currency may for a time be redeemable, but it constantly tends to become irredeemable.

We have been saved from inflation and an increased issue of irredeemable paper money only by the veto of a President, the policy of whose financial secretary had led logically and directly to the vicious legislation which was stopped by his veto.

Great Britain has its land question, we have the money question to be determined; both appalling in the consequences that may ensue from a false policy.

May not the record of history in both cases be the same, —that the principles of liberty and the sentiment of personal independence are so fully ingrained in the English race as to enable both branches to surmount the obstacles which their own legislators have placed in the way of their progress?

Whether the money be good or bad, whether the land be free or restricted, whether vested wrongs be sustained for a time, or vested rights promoted,—the sentiment of personal independence and individual liberty may be depended upon as the great safeguards of the English race, and will ultimately assure righteous laws.

In the first lecture which I gave you this year, I endeavored to picture to you the beneficent function of the railroad and the steamship, in assuring a good subsistence to the people of many lands and far-distant places.

In this I have treated the more abstract method by which distribution is promoted.

In the merely material work of the railroad, skill and intelligence only may suffice, but the conduct of the bank calls also for character and integrity of the highest order. In the history of commerce the great banker may, perhaps, stand first among those who have guided the great exchanges of the world, and who have made civilization possible.

EDWARD ATKINSON.

BROOKLINE, MASS., March, 1880.

THE RAILWAY, THE FARMER, AND THE PUBLIC

By EDWARD ATKINSON

[Reprinted from the *Manufacturers' Gazette* of Saturday, August 9, 1884]

THE RAILWAY, THE FARMER, AND THE PUBLIC.

I.

THE present condition of business, which may be called a partial commercial paralysis rather than an acute commercial crisis, the reduction in the prices of some of the most necessary articles of clothing and of food since 1882, the actual acute crisis in the stock-market, and the enormous reduction in the prices of railway securities, all alike point to subtle and powerful causes of change, perhaps of a permanent character, which cannot be explained by any superficial consideration of "corners," so called, or of the work of "bulls and bears," either in produce or in railway stocks or bonds. It is probably beyond the power of any investigator to make a complete analysis of all the forces which have produced these results. The utmost which can be done is to give a direction to thought and observation, leaving to the future to disclose the actual facts in all their bearings.

In pursuance of this great subject, let us first consider some of the most potent causes of permanent change in respect to the production and distribution of the necessary articles pertaining to the subsistence of the people, which have occurred since the end of the war. Food, clothing, and shelter are the subjects of primary consideration. Fuel is secondary in its application to household economy, but is of the first importance in the production of metals. With respect to food: Prior to the invention of the railroad and for a long period afterward—or until the railway service of the United States became finally and fully connected, East and West, which was about the year 1861,—the greater part, of the substantial food of each community was of necessity produced within a short distance of each town, city, or populous centre, owing to the necessary cost of distributing corn, meat, and dairy products in bulk by wagons. Under these conditions the best land in each State, or even in the separate sections of each State, near towns or cities, was of necessity devoted to the production of the coarser staples, *i. e.*, Indian corn, hay, meat, potatoes, and the like. The central parts of New York State and many parts of Pennsylvania were the sources of the greater part of the supply of wheat, but Western corn was unknown in Eastern markets. As distribution became less costly, especially after the final consolidation of the railway service in 1869, those coarser and more bulky products of agriculture became in a sense border or pioneer crops, and much land which had previously been devoted to their production in the East was now released and became used for market gardens, small fruits, and for other purposes. Central New York still produces as much wheat as ever, but a vast addition has been made of other salable crops, and agriculture is much more profitable than when wheat was the principal salable or money crop. The final consolidation of great railway systems took effect after the war, about the years 1869 and 1870, and in a treatise entitled "The Railway and the Farmer," published by the writer in 1881, he pictured in the graphical method the coincidence in the increase of the great grain crops of the country with the extension of the railway mileage. This coincident increase went on from 1865 to 1880, from over 1,100,000,000 to over 2,400,000,000 bushels, culminating in that year in the production of the largest grain crop ever before raised in the United States, and scarcely exceeded since.

Throughout this period there was a constant reduction in the charge for railway service, accompanied by a vast increase in the quantity of grain and other produce moved; but, measuring the prices by the gold standard, there was no substantial decrease in the price in the East of the principal farm products of the West. These facts will duly appear by the consideration of the graphical tables and the figures submitted herewith. Attention is especially called to the changes which occurred from 1869 to 1880 inclusive. It will be remembered that the year 1880, following the resumption of specie payments, was a year of great prosperity in every branch of production, whether in agriculture, mining, manufacturing, or in that part of the production or leading forth of useful commodities to the service of man which is commonly called distribution. All the work which is performed under either of these names is but a conversion of forces, *i. e.*, moving something from the soil or the mine for the use of man.

TABLE I.

GRAIN CROPS OF THE UNITED STATES.

Maize, Wheat, Rye, Oats, Barley, Buckwheat.

Year.	Bushels.
1865	1,127,499,187
1866	1,343,027,868
1867	1,329,729,400
1868	1,450,789,000
1869	1,491,412,100
1870	1,629,027,600
1871	1,528,776,100
1872	1,664,331,600
1873	1,538,892,891
1874	1,455,180,200
1875	2,032,235,300
1876	1,962,821,600
1877	2,178,934,646
1878	2,302,254,950
1879	2,434,884,541
1880	2,448,079,181
1881	2,066,029,570
1882	2,699,394,496
1883	2,623,319,089
1884	2,981,920,332

TABLE 2.

MILES OF RAILROAD IN OPERATION ON THE 1ST JANUARY IN EACH YEAR AND THE MILES ADDED IN THE YEAR ENSUING.

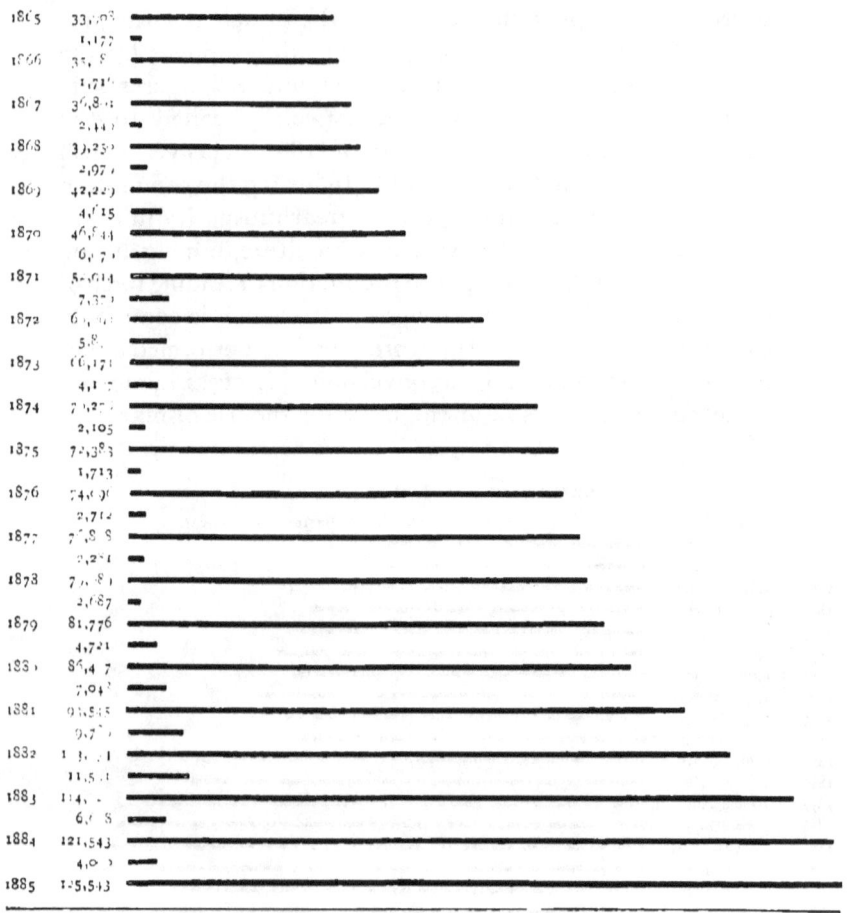

What then has happened since the year 1880? Railway mileage has increased since Jan. 1, 1880, over forty per cent. The crops of grain increased in 1882 ten per cent. as compared to 1880, and in 1883 a little over seven per cent., and yet these crops are more than ample to meet the present demand of the country; and since 1880 there has been first a rise and then a small reduction in the price of the leading farm products, as will appear by consideration of the graphical tables given herewith. The thirteen tons of beef, pork, wheat, corn oats, butter, wool, and lard which have been taken as the unit in this consideration, which were worth $632.68 in gold in 1869, $631.32 in 1880, $776.13 in 1882, were worth on June 15, 1884, $621.75. That these prices have been even so well maintained at this time gives proof of the continued prosperity of agriculture in spite of adversity elsewhere. The charge for moving these products on the principal railroads has fluctuated but little since 1879; it may be at this moment a little less than at that time, but if the charge is now less it is below the cost of the service and cannot be continued. Our great production of grain at less and less cost, and our great reduction in the charge for distribution, have been met since the year 1880 by increasing crops in other countries, coupled with improved methods of distribution, not, it is true, equal to our own, but yet working a possible future change in all the conditions of agriculture in this country so far as the wheat crop is concerned.

In the treatise upon "The Railroad and the Farmer" several computations were made as to the number of dollars which this reduction in the railway charge represented. It is something enormous. Had the actual quantity of merchandise moved by the railroad in the year 1880 been subjected to the average rate per ton per mile which was charged from 1866 to 1869 inclusive, the difference would have amounted to at least $500,000,000 and perhaps $800,000,000 more than the actual charge of 1880; and yet, up to this period, the prices of leading farm products had not been substantially affected by this enormous change,—that is to say, Eastern consumers of Western productions as yet received no benefit from this great reduction in the cost of distribution. But while consumers in the East may have as yet received little benefit in a direct reduction in the prices of Western produce, yet indirectly the benefit has been measureless. The grain and meat needed for a year's subsistence of one person, which would have cost a large portion of the time and labor to raise upon a comparatively sterile soil, to which agricultural machinery can be applied in least measure, is moved a thousand miles for a sum equal only to one day's wages of a common laborer. On the other hand, we import annually articles which are free of duty to the amount of $200,000,000 and one third of dutiable imports of the value of $150,000,000, which are either articles of food or crude materials which enter into all the processes of domestic industry, and these are all bought and paid for with the excess of grain, meat, and dairy products which we could not eat, the excess of cotton which we could not spin, the excess of oil which we could not burn, all of which would either be not produced or would be wasted if the low charges upon our railroads did not enable us to export them.

The consolidation and more effective service of the railways of the United States has been in the nature of a great and novel invention, and it has worked, as all great inventions work, for the time being, namely, to the immediate benefit of a relatively small part of the community,—that is to say, to the producers of particular substances. It is, perhaps, now working as other great inventions work in the secondary stage, namely, more to the benefit of the consumers. And yet even this is doubtful. The rapid increase in our home consumption seems to be sufficient to maintain prices, even when exports are greatly lessened. The world will, however, hereafter be less subject to local scarcity, less subject to particular famine; and a great mass of consumers of food may hereafter be required to devote a less proportion of their own labor to procuring the great staple articles of food. The forces in action in this matter have, therefore, been vastly greater than have appeared upon the surface, and a temporary retardation in the working of these forces by corners in grain and the like have been insignificant incidents of little permanent consequence.

Let us now consider the influence of these changes and of other great changes in their effect upon the railroads themselves. From a compilation of the statistics given in the census of 1880, coupled with a consideration of the data contained in Poor's Railway Manual, it is manifest that the staple articles of food—corn, meat, and dairy products—constitute, at least, fifty per cent. of the tonnage moved over all the railroads of the United States. They of course constitute a much larger proportion on some railroads than on others. Coal and timber in its various forms constitute not less than thirty per cent. of the remainder, and probably a yet larger proportion. If we reduce bushels to tons we find that the present average grain crop of the United States weighs 75,000,000 tons. Hay weighs from 30,000,000 to 35,000,000 tons; it is not all moved by the railway in its primary form, but if we add to the hay which is moved its product in the secondary form of meat and dairy products, we find a probable tonnage of 30,000,000 to 40,000,000 tons. It is more difficult to convert the timber products into tons, but approximately coal and timber together amount to over 100,000,000 tons. We therefore have over 200,000,000 net tons of food, fuel, and materials for shelter to be moved by a railway, at some point or in some part of their distribution, even if they have been moved part way by water on the road from producer to consumer. On the other hand, the entire production of metals within the limits of the United States is less than six million tons, cotton less than two million, wool less than half a million; and although these articles are converted into many different forms, and are moved twice, thrice, four times, or more, yet in the aggregate, after allowing for all duplications, they cannot amount to over twenty per cent., as compared to grain, timber, and coal, eighty per cent. From the census data and from the figures of Poor's Manual it would be difficult to make out over fifteen per cent. of miscellaneous merchandise in weight, consisting of metals, fibres, machinery, fabrics, and miscellaneous goods and wares, as compared to eighty-five per cent. in weight of food, fuel, timber, and other primary or crude products of the field, the coal mine, or the forest.

Now, then, if the grain, hay, and meat product—that is, the food of the people—constitutes one half the substance moved by the railway, and if this product has not increased in any measure beyond ten per cent. during the last four years, in which period the railway mileage has increased forty per cent., we have a sufficient explanation of all the disturbance in railway stocks and bonds. Moreover, a very large proportion of the railway construction from 1869 to 1880 inclusive represented a very much higher actual outlay or cost than the actual outlay or cost of what has been constructed since. The extreme example of this change is to be found in the reduction of the price of steel rails from over $150 a ton to less than $30 in gold, with a corresponding decrease in the cost of all the metal work pertaining to railways.

Now, it matters not how much may be the nominal amount of the stock and bonds issued either before or since 1880. It matters not whether a half or two thirds or three fourths even of any railroad is represented by what is called watered stock or not. All these enterprises are now brought face to face with the simple question—*Is there enough material to be moved, adjacent to their respective lines, at existing rates of freight, by which an income on actual cost can be earned, basing such cost upon what it would now be if the roads were constructed to-day?* It may be that watered stock, so called, which was issued before the great reduction in railway charges, may now be sustained by actual intrinsic value of double-track, equipment, or connections, since paid for out of earnings; or it may be, as in the case of the New York Central Railroad, that the right of way and terminal real estate is now worth a very large share, if not as much as all the outstanding stock and bonds; this does not alter the main question as above stated.

It will presently be made apparent that the charges for moving merchandise on long-established and fully equipped roads had been reduced in 1879 to the lowest possible terms consistent with even a small profit; therefore all new roads are met by one of three questions: First, if extensions into new sections, will the prices of possible products warrant the movement of crops except at rates which will barely sustain the road on a basis of cash cost? Second, if parallel roads, are they capable of being sustained at all? Third, if new roads in a section already well furnished, is there local traffic enough to pay even simple interest on a cash cost? In other words, have we not entered upon the final period in the history of railroads, to wit: the period in which they must be treated by their owners on a strictly commercial basis for the purpose only of earning a moderate income on the actual cash cost?

Before pursuing the subject further, with a view to considering the reasons why we may perhaps expect a speedy return of substantial prosperity after the railway system has become adjusted to these new conditions, I now submit certain tables which were originally constructed for an article on the "Railroad and the Farmer," published in 1881, which tables have been corrected and extended to the present date. I am indebted to the following authorities for the data on which these tables are based: The Department of Agriculture of the United States; E. H. Walker of the Produce Exchange of New York; Poor's Railway Manual; Messrs. Mauger and Avery of New York and Boston; G. R. Blanchard of New York; H. Sabine, Railroad Commissioner of Ohio; the reports of the Iron and Steel Association; and the United States Census of 1880.

The grain crops having increased only an average of five per cent., while the railway mileage increased more than forty, a part of which extension consisted of new routes from West to East, we may naturally look for a reduction of the tonnage on any principal route between West and East, and this we find even on the Lake Shore and New York Central, as will appear by tables 3 and 4.

TABLE 3.

LAKE SHORE & MICHIGAN SOUTHERN RAILROAD.—ACTUAL TONS MOVED.

Yr.	Miles.	Tons Moved.	Increase of Tons Moved Consolidated in this year.
'69			
'70	1,013	2,978,725	
'71	1,073	3,784,525	
'72	1,136	4,443,092	
'73	1,154	5,176,661	
'74	1,175	5,221,267	
'75	1,176	5,022,490	
'76	1,177	5,635,167	
'77	1,177	5,513,398	
'78	1,177	6,098,445	
'79	1,177	7,541,294	
'80	1,177	8,350,336	
'81	1,177	9,164,508	
'82	1,274	9,195,528	
'83	1,340	8,478,605	

LAKE SHORE & MICHIGAN SOUTHERN.—TONS MOVED ONE MILE.

Year.	Tons Moved one Mile.	Increase of Traffic, Tons per Mile.
1870	574,035,571	
1871	733,670,696	
1872	924,844,140	
1873	1,053,927,180	
1874	999,342,041	
1875	943,236,161	
1876	1,133,834,828	
1877	1,080,005,561	
1878	1,340,467,826	
1879	1,733,423,440	
1880	1,851,166,018	
1881	2,021,775,468	
1882	1,892,868,224	
1883	1,689,512,415	

LAKE SHORE & MICHIGAN SOUTHERN.—CHARGE PER TON PER MILE.
AVERAGE UPON ALL CLASSES OF MERCHANDISE.

Year.	Freight Receipts. DOLS.	Charge. CTS.	Decrease of Charge per Mile.
1870	8,746,126	1.504	
1871	10,341,218	1.391	
1872	12,824,862	1.374	
1873	14,192,369	1.335	
1874	11,918,350	1.180	
1875	9,639,038	1.010	
1876	9,405,629	.870	
1877	9,476,608	.864	
1878	10,048,952	.734	
1879	11,288,261	.642	
1880	14,077,294	.750	
1881	12,659,787	.617	
1882	12,023,577	.62	
1883	12,400,094	728	

TABLE 4.

NEW YORK CENTRAL & HUDSON RIVER RAILROAD.—ACTUAL TONS MOVED.

Yr.	Miles.	Tons Moved.
'69	842	3,190,840
'70	842	4,122,000
'71	844	4,532,056
'72	850	4,303,905
'73	858	5,522,724
'74	1,000	6,114,678
'75	1,000	6,001,984
'76	1,000	6,803,680
'77	1,000	6,351,356
'78	1,000	7,635,413
'79	1,000	9,045,753
'80	1,000	10,533,038
'81	993	11,591,370
'82	993	11,330,393
'83	993	10,892,440

NEW YORK CENTRAL & HUDSON RIVER RAILROAD.—TONS MOVED ONE MILE.

Year	Tons Moved One Mile
1869	589,362,849
1870	763,087,777
1871	888,327,865
1872	1,020,908,885
1873	1,246,650,063
1874	1,311,360,707
1875	1,414,008,729
1876	1,671,447,055
1877	1,619,948,685
1878	2,048,755,132
1879	2,295,827,387
1880	2,525,139,145
1881	2,646,834,095
1882	2,394,749,310
1883	2,200,896,780

NEW YORK CENTRAL & HUDSON RIVER RAILROAD.—CHARGE PER TON PER MILE.

AVERAGE ON ALL CLASSES OF MERCHANDISE.

Year	Receipts. DOLS.	Charge. CTS.	Decrease of Charge per Mile.
1869	14,066,386	2.387	
1870	14,327,418	1.853	
1871	14,647,580	1.649	
1872	16,259,650	1.592	
1873	19,616,018	1.573	
1874	20,348,725	1.462	
1875	17,899,702	1.275	
1876	17,573,065	1.051	
1877	16,424,317	1.014	
1878	19,045,830	.920	
1879	18,270,250	.776	
1880	22,199,966	.879	
1881	20,736,750	.783	
1882	17,672,252	.738	
1883	20,142,433	.910	

It will be observed that so long as the increase of crops kept pace with the increase of railroads, and both were accompanied by such an export demand for breadstuffs as to maintain the through traffic, the rate of charge diminished, but when the traffic diminished the rate of charge soon began to show a slight increase. This is, doubtless, caused by the change in or less proportion of through traffic. The following table shows that while the traffic on the New York Central and Lake Shore decreased in some measure in 1882 and 1883, yet the traffic on all the roads reporting in New York increased. The data hereafter given from the statistics of Ohio, in which the through and local tonnage are separated, also fully sustain this view, and show how railroads which may at first be mainly supported by through traffic are ultimately supported mainly by local traffic. Table 5 shows the continued increase of traffic on all railroads reporting in the State of New York.

This table includes some roads of which only a small part actually lies within the limits of the State.

The following table, No. 6, gives the earnings, expenses, and profits per ton per mile on the New York Central and Hudson River Railroad in 1855, 1865, and from 1869 to 1883 inclusive.

It will be manifest that when such a strong and rich corporation as this has been forced to do its work for the last five years at a profit of less than a quarter of a cent per ton per mile, or *one fortieth of a cent profit for moving a barrel of flour one mile*, there is no margin for any further reduction of any moment; and it also becomes apparent that the construction of a parallel line for the purpose of sharing this work was a pure waste of capital and almost wholly a loss to the purchasers of the securities, and that the ruin of its promoters might have been foretold at the beginning, as it was

TABLE 5.

TONS MOVED UPON ALL THE RAILROADS REPORTING IN THE STATE OF NEW YORK.

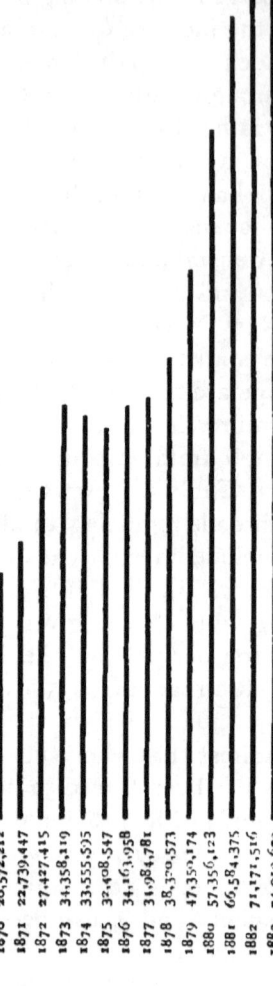

Year	Tons
1870	20,572,212
1871	22,739,447
1872	27,427,415
1873	34,358,119
1874	33,555,595
1875	32,408,547
1876	34,163,958
1877	31,984,781
1878	38,370,573
1879	47,359,174
1880	57,356,123
1881	66,584,375
1882	71,171,516
1883	74,019,679

THE RAILWAY, THE FARMER, AND THE PUBLIC. 245

TABLE 6.
NEW YORK CENTRAL AND HUDSON RIVER RAILROAD.

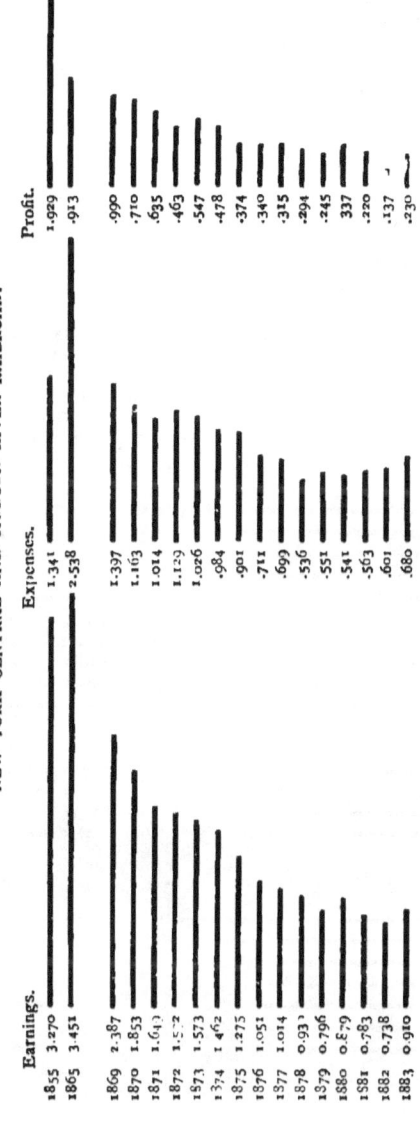

	Earnings.	Expenses.	Profit.
1855	3.270	1.341	1.929
1865	3.451	2.538	.913
1869	2.387	1.397	.990
1870	1.853	1.163	.710
1871	1.649	1.014	.635
1872	1.572	1.129	.463
1873	1.573	1.026	.547
1874	1.462	.984	.478
1875	1.275	.901	.374
1876	1.051	.711	.340
1877	1.014	.699	.315
1878	0.930	.536	.294
1879	0.796	.551	.245
1880	0.879	.541	.337
1881	0.783	.563	.220
1882	0.738	.601	.137
1883	0.910	.680	.230

by more than one observer. By whom this great reduction in freight charges has been mainly or directly enjoyed will appear from the following computation of the value of thirteen tons of staple produce, and the comparison of the freight charge thereon. It must, however, be remembered that the greater reduction has been made on the through traffic on grain and provisions than on any other class of traffic, hence the tables do not show the full benefit to the Western producers.

TABLE 7.

COST OF 20 BARRELS OF FLOUR, 10 BEEF, 10 PORK, 100 BUSHELS WHEAT, 100 CORN, 100 OATS, 100 POUNDS BUTTER, 100 LARD, AND 100 FLEECE WOOL, IN NEW YORK CITY, AT THE AVERAGE OF EACH YEAR, COMPILED BY MONTHS, IN CURRENCY AND GOLD; COMPARED GRAPHICALLY WITH THE DECREASE IN THE CHARGE PER TON PER MILE ON THE NEW YORK CENTRAL & HUDSON RIVER RAILROAD, DURING THE SAME PERIOD.

Year	Cost in Currency.	In Gold.
1869	$845.58	$632.68
1870	891.80	776.02
1871	821.60	735.33
1872	760.24	675.92
1873	755.68	662.50
1874	831.98	748.54
1875	800.28	696.40
1876	727.4	651.74
1877	780.29	751.95
1878	575.41	569.81
1879	568.34	568.34
1880	631.32	631.32
1881	703.10	703.10
1882	776.13	776.13
1883	662.11	662.11
1884	621.75 [June]	621.75

Year	Decrease in the charge per Ton per Mile, N. Y. C. & H. R. R. R.—In Currency.	Decrease in the charge per Ton per Mile, N. Y. C. & H. R. R. R. In Gold.
1869	2.38 cts.	1.78 cts.
1870	1.85 "	1.64 "
1871	1.65 "	1.40 "
1872	1.50 "	1.41 "
1873	1.57 "	1.38 "
1874	1.46 "	1.31 "
1875	1.27 "	1.11 "

1876 1.05 cts. ▄▄▄▄▄▄▄▄▄▄▄   .94 cts. ▄▄▄▄▄▄▄▄▄
1877 1.02 "   ▄▄▄▄▄▄▄▄▄▄   .97 "   ▄▄▄▄▄▄▄▄▄▄
1878   .93 "   ▄▄▄▄▄▄▄▄▄   .92 "   ▄▄▄▄▄▄▄▄▄
1879   .79 "   ▄▄▄▄▄▄▄▄   .79 "   ▄▄▄▄▄▄▄▄
1880   .88 "   ▄▄▄▄▄▄▄▄▄   .88 "   ▄▄▄▄▄▄▄▄▄
1881   .78 "   ▄▄▄▄▄▄▄▄   .78 "   ▄▄▄▄▄▄▄▄
1882   .73 "   ▄▄▄▄▄▄▄   .73 "   ▄▄▄▄▄▄▄
1883   .91 "   ▄▄▄▄▄▄▄▄▄   .91 "   ▄▄▄▄▄▄▄▄▄

Freight charge in year 1855, in
  gold, 3.27 cts. . . . .   ▄▄▄▄▄▄▄▄▄▄▄▄▄▄▄▄▄▄▄▄▄▄▄▄▄▄▄▄▄▄▄
Freight charge in year 1865, in
  currency, 3.45 cts. . . . . .   ▄▄▄▄▄▄▄▄▄▄▄▄▄▄▄▄▄▄▄▄▄▄▄▄▄▄▄▄▄▄▄▄

To whom the advantage has accrued will be made yet more clear by setting off the actual dollars of freight charges on thirteen tons moved 1,000 miles, or from Chicago to New York or Boston, at the average rates charged by the New York Central & Hudson River Railroad on all classes of traffic from 1869 to 1883, inclusive, using gold values only in respect to prices and rates.

TABLE 8.

PRICES IN GOLD IN THE NEW YORK MARKET OF 20 BARRELS FLOUR, EXTRA STATE, 100 BUSHELS WHEAT, MILWAUKEE CLUB, 100 BUSHELS CORN, WESTERN MIXED, 100 BUSHELS OATS, 10 BARRELS MESS PORK, 10 BARRELS MESS BEEF, 100 POUNDS LARD, 100 POUNDS STATE DAIRY BUTTER, 100 POUNDS MEDIUM WASHED CLOTHING WOOL, COMPARED WITH CHARGE REDUCED TO GOLD OF MOVING THE ABOVE QUANTITY, EQUAL TO 13 TONS 1,000 MILES, AT THE AVERAGE RATES CHARGED BY NEW YORK CENTRAL & HUDSON RIVER RAILROAD, 1869 TO 1883, INCLUSIVE.

Year	Cost In Gold	Prices.	Decrease in the Charge per Ton per Mile, N. Y. C. & H. R. R.—In Gold.	
1869	$662.68	▄▄▄▄▄▄▄▄▄▄▄▄▄▄	1.78 cts.	▄▄▄▄▄▄▄▄▄▄▄▄▄▄▄▄▄
1870	775.02	▄▄▄▄▄▄▄▄▄▄▄▄▄▄▄▄	1.64 "	▄▄▄▄▄▄▄▄▄▄▄▄▄▄▄▄
1871	735.33	▄▄▄▄▄▄▄▄▄▄▄▄▄▄▄	1.40 "	▄▄▄▄▄▄▄▄▄▄▄▄▄▄
1872	675.92	▄▄▄▄▄▄▄▄▄▄▄▄▄▄	1.41 "	▄▄▄▄▄▄▄▄▄▄▄▄▄▄
1873	662.50	▄▄▄▄▄▄▄▄▄▄▄▄▄▄	1.38 "	▄▄▄▄▄▄▄▄▄▄▄▄▄
1874	748.54	▄▄▄▄▄▄▄▄▄▄▄▄▄▄▄▄	1.31 "	▄▄▄▄▄▄▄▄▄▄▄▄▄
1875	696.40	▄▄▄▄▄▄▄▄▄▄▄▄▄▄▄	1.11 "	▄▄▄▄▄▄▄▄▄▄▄
1876	651.74	▄▄▄▄▄▄▄▄▄▄▄▄▄▄	.94 "	▄▄▄▄▄▄▄▄▄
1877	751.95	▄▄▄▄▄▄▄▄▄▄▄▄▄▄▄▄	.97 "	▄▄▄▄▄▄▄▄▄▄
1878	569.81	▄▄▄▄▄▄▄▄▄▄▄▄	.92 "	▄▄▄▄▄▄▄▄▄
1879	568.34	▄▄▄▄▄▄▄▄▄▄▄▄	.79 "	▄▄▄▄▄▄▄▄
1880	631.32	▄▄▄▄▄▄▄▄▄▄▄▄▄	.88 "	▄▄▄▄▄▄▄▄▄
1881	703.10	▄▄▄▄▄▄▄▄▄▄▄▄▄▄▄	.78 "	▄▄▄▄▄▄▄▄
1882	776.13	▄▄▄▄▄▄▄▄▄▄▄▄▄▄▄▄	.73 "	▄▄▄▄▄▄▄
1883	662.11	▄▄▄▄▄▄▄▄▄▄▄▄▄▄	.91 "	▄▄▄▄▄▄▄▄▄
1884	621.75	▄▄▄▄▄▄▄▄▄▄▄▄▄ [June]	.83 "	▄▄▄▄▄▄▄▄

Year	Dollars, 13 Tons, 1,000 Miles.	Per cent. of Freight Charge to Value in New York.
1869	231.40	36.61
1870	213.20	27.47
1871	182.00	24.76
1872	183.30	27.16
1873	179.40	27.05
1874	170.30	22.73
1875	144.30	20.73
1876	122.20	18.74
1877	126.10	16.76
1878	119.60	20.98
1879	102.70	18.08
1880	114.40	18.12
1881	107.40	15.27
1882	94.90	12.22
1883	118.30	17.87

The above proportions of the value of the produce absorbed by the freight charge should be reduced in just the measure that the rates per mile on the movement of grain and meat have been less than the average charge on the whole traffic. For instance, thirteen tons of grain have been brought from Chicago to New York at a lower charge by far than any of the above figures. This change would reduce the proportion of the charge now in greater measure than in the earlier part of the period under consideration.

But it may be said all these data are limited to the through traffic, and the local traffic is still subjected to onerous charges and unjust discrimination. In reply to which I submit Table No. 9, in which the receipts, expenses, and profits of all the railroads reporting in New York are analyzed and compared, by which it will appear that in 1879 the profit on all the traffic was brought down to less than one quarter of a cent per ton per mile, and has averaged less than that rate ever since. What it may be this year cannot yet be stated.

THE RAILWAY, THE FARMER, AND THE PUBLIC. 249

TABLE 9.

RECEIPTS, EXPENSES, AND PROFITS OF ALL THE RAILROADS IN NEW YORK, INCLUDING SEVERAL WHICH, LIKE THE LAKE SHORE AND BOSTON AND ALBANY, HAVE BUT A SMALL PORTION OF THEIR MILEAGE IN THE STATE.

Year.	Receipts per Ton.	Expenses.	Profits.
1870	1.7016	1.1471	.5545
1871	1.7005	1.1450	.5555
1872	1.6645	1.1490	.5155
1873	1.6000	1.0864	.5136
1874	1.4480	.9730	.4750
1875	1.3039	.9587	.3452
1876	1.1604	.8561	.3043
1877	1.0590	.7740	.2850
1878	.9934	.6940	.3094
1879	.8482	.5847	.2295
1880	.9220	.6030	.3190
1881	.8390	.5880	.2510
1882	.8176	.6010	.2166
1883	.8990	.6490	.2500

Tons moved in 1883 74,919,679
Tons one mile, 1883 9,286,216,628
Earnings . . . . . $83,464,919.

TABLE 10.

TONS MOVED ONE MILE—LOCAL AND THROUGH TRAFFIC, IN OHIO.

Year	Tons
1869	1,332,597,931
1870	1,673,017,563
1871	1,773,931,495
1872	2,923,232,084
1873	3,420,883,453
1874	3,717,622,979
1875	3,431,745,707
1876	3,799,377,649
1877	4,146,926,316
1878	4,286,378,572
1879	4,914,503,869
1880	6,655,562,182
1881	7,6.7,215,616
1882	8,244,019,566
1883	8,577,357,803

TABLE 11.

Year	Local.	Year	Through.
1869	7,599,105	1869	6,569,503
1870	7,347,083	1870	5,938,902
1871	8,374,159	1871	6,281,364
1872	12,62,392	1872	6,100,707
1873	19,384,932	1873	7,204,583
1874	18,374,756	1874	7,824,629
1875	15,493,697	1875	6,544,583
1876	14,648,719	1876	7,075,631
1877	16,755,546	1877	9,638,613
1878	16,059,219	1878	10,931,055
1879	17,152,825	1879	12,983,884
1880	25,797,025	1880	18,554,747
1881	29,531,212	1881	20,443,230
1882	30,474,136	1882	21,752,179
1883	42,308,712	1883	21,314,711

Had the rate of 1870 been charged on the traffic of 1883, the sum would have been at 1.7016 on 9,286,216,628 tons, carried one mile, $158,014,262; the actual charge was $83,464,919, making a difference of $74,549,343 saved on one year's traffic on the lines reporting in New York.

But again, let us examine the traffic of the great State of Ohio, midway between the grain fields of the far West and the manufacturing States of the far East, a State in which agriculture, mining, and manufacturing are combined. The invaluable tables of her Railroad Commisioner, Mr. Sabine, separate the through from the local traffic, and these tables show how, in the course of time, all our existing railway lines, except such speculative absurdities as those which have been built close alongside other tracks, may become self-sustaining and profitable.

Again we find that the average freight charge has been reduced to a little less than one cent per ton per mile, and there it has substantially rested for seven years, because it cannot go lower without stopping the traffic altogether.

TABLE 12.

Freight Charge.

Year	Charge
1869	2.446
1870	1.993
1871	2.215
1872	1.569
1873	1.566
1874	1.334
1875	1.259
1876	1.117
1877	.933
1878	.961
1879	.815
1880	.895
1881	.915
1882	.807
1883	.875

Tons per mile in 1883, 8,577,357,803, at 1869 rates, 2.446 . $201,800,000
At actual rate of 1883, .875 67,000,000

Difference . . . $134,800,000

At currency rate of 1869 reduced to gold, 1.80 . . . . $156,400,000
Actual 67,000,000

Difference $89,400,000
Difference on local traffic only . . $60,000,000

And once more we prove that, had the freight charge of 1869, reduced to gold, been put upon the traffic of 1883, the results would have been, in round figures, as follows:

8,577,357,803 tons moved 1 mile, at 1.80 cts. per ton per mile . $156,400,000
Actual charge 67,000,000

Difference $89,400,000

As two thirds of this was on local traffic the people of Ohio saved, in the single year 1883, $60,000,000 on their internal exchanges only. The ton mileage of New York and Ohio combined, in 1883, was as follows on all the roads reporting in each State:

New York, tons 1 mile 9,286,216,628
Ohio, " " 8,577,357,803

Total 17,863,574,431

Saving in New York as compared to gold rate of 1870 . . $74,549,343
Saving in Ohio, as compared to gold rate of 1869 . . . 89,400,000

Total $163,949,343

The reports of these two States covered about four tenths of the total ton mileage of the whole country in 1883, which was 42,361,068,260 tons carried one mile at a charge of $549,339,736. As great a reduction, or even greater, has been made on all roads which were in existence from 1866 to 1870, while the new roads have worked a yet greater saving, because they take the place of traffic by wagons or by rivers. At the ratable difference made on the New York and Ohio railways, the traffic of the whole country in 1883, which was done for the sum of $549,339,736, would have cost $950,000,000, or $400,000,000 more.

As we go back to the yet higher rates of 1866, '67, and '68, the difference rises to $600,000,000; if we compare the gold rates now with the currency rates then, the difference is yet more—even more than $800,000,000.

Again, I call attention to this as the true source of our increased power of subsistence—as the main source of our actual increase in capital, and also the source whence has come the fund for railway construction. Only a small part of this fund has been wasted; the speculative enterprises by which parallel lines have been built too near to existing lines ever to be of any value are limited to a few which any one, who is familiar with names, can identify. By far the larger portion, even of the forty-per-cent. extension in four years' time, will ultimately justify their existence, and will be sustained with a moderate income on a cash cost; but the day of profit on two, three, or four dollars of security issued for one paid in, has passed, let us hope, forever.

These tables could be extended to almost any extent; all the great lines show identical results, but it would be useless to multiply proofs. Suffice it, that whatever may have been the intention of the promoters of these enterprises, and to whatever extent they may have misled the investors who have risked their money in order to gain speculative profits —whatever proportion of the securities may be cash or water, —the competition not only of railway with railway but also of product with product, has forced the charge for transportation to the lowest point consistent with any profit whatever, even on the strongest lines and on those which have been called the greatest monopolies. These dry and voluminous statistics are presented with an assurance that they will be honestly considered, and will lead men to beware of meddlesome legislation affecting the most beneficent force by which a good subsistence is made common to all at the least cost.

In what follows I shall be obliged to repeat some of the data already given, in order to sustain the distinct and separate purpose of the remainder of this essay.

In connection with the foregoing figures I have been asked to give my views of the future immediate prospects of business in this country. What are such views worth? For any immediate application, absolutely nothing. Ask the apple-woman what apples will be worth next autumn, and her views may be worth as much for any immediate application as those of the most sagacious banker or merchant in the city. But a few facts may be given which have a bearing on the course of business during the next five years. I will give them for what they are worth, and every reader can draw his own deductions from them. Such facts may only be of use in estimating industrial forces covering a considerable period. The facts which caused the panic of 1873 were just as apparent in 1870 as they were during its action, but its exact date could not be foreseen. The long period of necessary depression, while the depreciation of the currency was being corrected, could be as clearly apprehended before 1873 as it could be during its continuance until 1879. The "boom" of 1880 was an obvious necessity, and was easily predicted in 1878 and '79. The commercial "paralysis" of 1883, and the railway panic ensuing in 1884, were both apparent and were foretold in the winter of 1881, although no date could be established in advance. With equal certainty the commercial activity of the near future, and the exceeding prosperity which must ensue, may be predicated on existing conditions, were it not for two uncertain factors. These are:

First, The silver question.

Second, Uncertainty in regard to the future financial policy of the Government.

In respect to the first there is still time to prevent the debasement of the standard of value to the level of a dollar of light weight, worth but little more than eighty cents in gold; but every year's delay will bring the country nearer to the inevitable disaster which must ensue from our existing acts of legal tender and coinage.

In respect to the second danger a few months will tell; in the meantime, *constructive enterprise* will wait the decision of the people as to whether their policy shall be one of peace, prosperity, reduced taxation, and recuperation; or one of uncertainty, probable aggression, possible war, and of the perversion of the functions of government to purposes of personal ambition and private gain. What effect a temporary cessation of constructive enterprise exerts will be fully treated hereafter. Assuming that both these special causes of disaster, of want of confidence, and of continued depression may be avoided, a period of great future prosperity may be predicated on present conditions, although no man can tell when the exact turn of the tide will come.

In order to comprehend the present conditions under which this country is now making greater real progress in material welfare than at almost any previous period in its history, certain elements must be considered in their relative proportions; for this purpose some figures of the census may be used. In respect to these figures it must be premised that the valuation of farms is probably under-estimated, that the capital in railways included the "water" which is now being squeezed out, and that the capital in manufactories was probably over-estimated. In considering the relation of proportion which these great branches of industry bear to each other we may therefore assume:

First, That the proportion of the national capital in improved lands and farm buildings, *i. e.*, in the instrumentality of primary production, is herein stated too low.

Second, That the capital in manufacturing, *i. e.*, in the instrumentality of conversion of crude materials into finished goods, and the capital in railways, *i. e.*, in the chief instrumentality for distribution, are herein stated too high ; but that the figures of the census fairly represent their relation or proportion to each other.

Omitting fractions, the respective capitals in these three great departments of industry were, in 1880, as follows, as given in the census :

Farm lands and farm buildings	$10,200,000,000
Railways	5,200,000,000
Manufacturing (listed under 332 different heads) .	3,000,000,000

Graphically represented, the relative proportion of these capitals is as follows :

Farms	───────────────────────────
Railways	───────────
Manufacturing	──────

It will be observed that the valuation of the farms includes the land; if we separate farm buildings, machinery, tools, and appliances from land—that is to say, separate all the actual capital upon the farm from the land, and add this sum to the capital in manufactures, the total productive capital, in both agriculture and manufactures, was about the same, perhaps a little more, than the single capital in railways. This brings into the clearest light the relative importance of distribution. In this country there is always enough for all, but where is it ? Our productive capacity is unlimited, and the main question is one of distribution. The railroad has solved a part of this problem, but there are more complex questions yet to be solved. It costs a third of the price of a baker's loaf to get a loaf of bread away from the oven, after it is baked, to the mouth of the consumer. [See Appendix I.]

Such being the relative proportions of capital in farming, manufacturing, and railways in 1880, what changes have occurred since, by which we may in part account for the great variations in the market value of either class of the above property? especially the reduction in the nominal value of railway property? How can we account for the railway panic, for the great private losses, and for the redistribution of property in railways, which is now going on? How can we account for the comparative stability in the value of manufacturing property of all kinds, and for the relative and actual prosperity of agriculture, the latter the most important factor of all in the condition of the country— the one great fact on which we may forecast future prosperity?

The reasons are not far to seek. At the foundation of agriculture lies the grain crop. Grain, or its secondary products, meat and dairy products, constitute the principal elements in weight of the tonnage of the railways. The average grain crop, from 1866 to 1869, inclusive, was 1,400,- 000,000 bushels. The average railway mileage from 1866 to 1869, inclusive, was 39,000 miles. The average grain crop from 1877 to 1880, inclusive, was 2,341,000,000 bushels. The average railway mileage from 1877 to 1880, inclusive, was 83,000 miles. The average grain crop of 1881 to 1883, inclusive, was 2,450,000,000 bushels. But our railway mileage is now, or was on the 1st of January, 1884, over 121,000 miles. What are the necessary conclusions from these figures?

From 1866 to 1880 one line after another was added to the great through lines from East to West; slowly but surely, down to 1880, the railway mileage gained a little upon the grain crop, the slight excess representing nothing more than the necessary cross-roads and side-lines. The markets of the world also kept even pace, increasing supply of grain and meat was met by increasing demand, down to 1880 inclusive. In 1869 thirteen tons of produce, already listed, were worth in gold in the city of New York, $632.68; in 1880 the same quantities were worth $631.32. But in 1880 the increase of demand culminated; exports have fallen off about as fast as the home demand has increased, yet the same quantities of the same articles are worth at this time (June 15, 1884) $621.75, or less than two per cent. reduction. Observe, however, with only five per cent. increase in the average crop of grain since 1880, we have more than forty per cent. increase in the railway mileage in the last four 'years. We *have two through lines where one is needed*, and the end of speculative construction is therefore plainly to be seen. We have passed through the period of railroad inception and of detached sections or lines, through the period of consolidation, through the period of needed extension, through the period of the speculative promotion of useless parallel lines by means of construction companies; and we have *now* at last reached the period of adjustment to wholesome conditions and of construction limited to the necessity for cross-roads, side-lines, and special or local roads for the use of small districts. Even this latter need will probably require this year 4,000, afterward 5,000 to 6,000 miles, to be added to our mileage every year.

But while this vast extension in railway mileage has been in progress, the freight charges on all the railways of the country, and especially on the through lines, were reduced between 1869 and 1880 *two thirds;* that is to say, the charge on the thirteen tons carried from West to East 1,000 miles, which was over $180 in 1869, was less than $60 in 1880, and has since fluctuated but little, sometimes a little less, sometimes a little more. In fact, there can hereafter be no further great reduction in freight charges. The bottom was reached in 1880; the entire profit on the whole tonnage of the New York Central and Hudson River Railroad, in 1882, was but a trifle over one eighth of a cent a ton per mile. In exact figures it was .1370 cents per ton per mile. That is to say, at the average rate of profit on the whole traffic, grain and flour being carried at much lower relative rates, the actual profit in 1882 for moving a barrel of flour 1,000 miles, or from Chicago to New York, was *thirteen cents*, or about one third part of the cost of the barrel in which the flour is packed. In 1880 the possibility of any further permanent reduction on established lines of railway therefore ended, until some new invention shall reduce the cost of the service. So far as parallel or competing main lines have been constructed since that date the capital expended has been utterly wasted.

The elimination of what has been called " watered stock and bonds," which cannot affect the charge for transportation in any manner, is, therefore, in process of accomplishment by methods far more potent than any possible legislative acts, namely, by the triple competition to which railways are subjected: First, The competition of waterways; Second, The competition of one railway with another; Third, The competition of product with product in the great markets of the world. The charge which can be put upon the wheat of Dakota or Iowa for moving it to market is fixed by the price at which East Indian wheat can be sold in Market Lane. The railway mileage Jan. 1, 1880 (when the possibility of any further reduction in freight charges covering any profit whatever commensurate with a fair but very low revenue was practically reached), was 86,-497, represented by over $5,000,000,000 of securities. Jan. 1, 1884, it was 121,542 miles, represented by over $7,000,-

000,000 of securities. Since Jan. 1, 1880, we have increased our population *twelve and a half* to *fourteen* per cent., our grain crops *five* per cent., our railway mileage *forty* per cent. Aside from grain, the increased production of other commodities has probably not averaged a greater rate than the increase in population.

Having, therefore, reached the end of construction companies, of speculative building, and of the issue of two, three, or four dollars of security for one dollar actually paid, we are now entering upon a period of railway adjustment; that is to say, of earnings limited to a moderate rate of possible dividend on what the needed portion of the present railroad mileage would cost at the present actual prices of labor and materials, unnecessary parallel roads being deprived of all earning capacity.

How much nominal property will be wiped out of existence, and how many individuals will suffer, it matters not, except to the sufferers. Hereafter, the people of the United States will be served by 120,000 miles of railway, operated at the lowest possible cost, and which will be extended only so fast as prudence or necessity may require. This service, which would have cost producers or consumers $1,000,000,000 gold to $1,350,000,000 currency a year at the rates which were charged from 1866 to 1869 inclusive, is now performed for about $550,000,000 per year, a saving of $450,000,000 to $800,000,000 per year.

While this revolution has been accomplished, the leading farm products, of which I have given a list, thirteen tons in all, which were worth in gold coin in the city of New York in 1869 $632.68, and which are now worth $621.75, have averaged during the whole series of years $679.50. In this period of so-called depression and disaster it therefore appears that the prices of the staple products of our Western farms and of our Eastern dairies in New York are but eight per cent. lower than the average of the last fifteen years. Upon the misfortunes of railway owners we may, therefore, predicate the past and present and also the future prosperity of the farmer; and upon the prosperity of the farmer we may also assume the future prosperity of the manufacturer, because their interests are identical. Another fact must also be considered: during the period under consideration the mechanism of distribution has not only been increased in this wonderful, measure, accompanied by the vast increase of crops, but the increase of crop has been much greater than the increase of population. In 1869 the production of grain was about forty bushels per capita, in 1884 it was more than fifty-two bushels, an increase of thirty per cent. If the general product of agriculture may be represented by the grain crop, if follows that where the result of farm labor worth \$632.68 in 1869 represented a given amount of labor, in 1884 it represented only about two thirds as much.

At the risk of unnecessary repetition, let me again call attention to the salient facts in respect to the State of Ohio which are disclosed in the admirable reports of her railway commissioner. I again call attention to these points, because from them the restoration of value to many lines of railroad now embarrassed, may be implied. This State lies midway between East and West. In 1883 it contained 6,897 miles of railroad, against 3,324 in 1869. In 1869, the actual tons moved over all the railways reporting in the State numbered 14,559,704, of which fifty-five per cent. represented local traffic and forty-five per cent. through traffic. In 1883, 63,683,423 tons were moved, of which sixty-six and one-half per cent. represented local traffic and only thirty-three and one-half per cent. through traffic, showing how the local traffic gains, both absolutely and relatively. The charge per ton per mile in 1869 was 2.446 cents; in 1883, only .875 cents per ton per mile. Graphically the Ohio railroad traffic may be represented in this way:

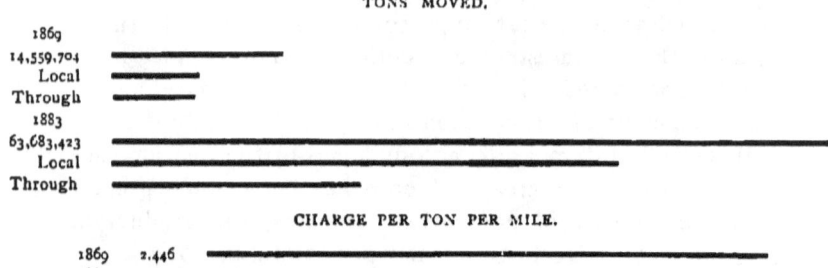

The actual freight charge on all the railroads reporting in Ohio in 1883 was, in round figures, $67,000,000. Had this traffic been subjected to the charge of 1869 the sum would have been $201,800,000.

The difference between these two sums is, in currency, $134,800,000; in gold, $89,400,000. Now since two thirds of this traffic was *local* traffic, the saving in rates to the people of Ohio since 1869, on their local traffic only, was, in currency, $90,000,000; in gold, $60,000,000.

This is the difference on the work done in a single year in a single State! The commissioner may well say in his letter to me transmitting this information, "I am glad to say that in Ohio the people and the railways are at peace." The example of Ohio is a crucial instance of how the railways have diversified the employment of the people, and how this very diversity afterward sustains the railways, as the local traffic steadily increases in its relative proportion. If such has been the gain in a single State—$60,000,000 saved on the local traffic of a single year, as compared to the rate of only fourteen years since—the secret of increasing wealth, lower rates of interest on capital and increasing wages to the laborer, is not far to seek. Nor can depression or adversity long hold a place under such conditions, given only stability to the standard of value, judicious reduction of taxes, freedom from an aggressive foreign policy, and the limitations of legislative action to assuring the publicity of the accounts of public corporations, without futile attempts to control them.

But to return to the adjustment of the value of railway property. Stocks and bonds, nominally representing $7,000,000,000 worth of property, apparently depreciated at least $1,500,000,000 within the year 1884, or, in other words perhaps $1,000,000,000 of " water " was squeezed out, and during the process the true value of the remainder has been temporarily depressed $500,000,000, from which depression it must soon recover. Such vast changes, of which the conspicuous frauds of a few persons are but the surface indications (the greatest knaves not having even yet been ruined), could not fail to affect in a most profound degree all banks and other institutions of credit, and in less measure all who engaged in production and distribution. And yet no tradesman, no merchant, no manufacturer has yet failed who had not long been insolvent, and hardly a banker ; the prices of the staple farm products are almost the same as in 1869 and in 1880, the latter a year of great prosperity, and there have been few manufacturing accounts made up which showed an actual loss, while many branches of business are prosperous. There is no safer barometer than the production and consumption of iron. This branch of industry has been said to be more depressed than almost any other ; but what are the facts? According to the records of the Secretary of the Iron and Steel Association, the production of pig-iron in the last five years, which included the "boom" year 1879, was as follows:

NET TONS.

1879	3,070,875
1880	4,295,414
1881	4,641,564
1882	5,178,122
1883	5,146,972

Let it be observed that in the face of lower and lower prices, from 1880 to the present time, and in spite of a reduction of nearly one half in the rails laid upon new railroads in 1883 as compared to 1882, the production and consumption of American iron in the year of so-called greatest depression, 1883, was sixty-six per cent. greater than in the "booming" year, 1879. The consumers of iron may well be satisfied with the construction of modern furnaces well placed, in which iron is made at so much lower cost that, in the face of eleven-per-cent. reduction in the price of 1880, the production of the metal which is at the foundation of all arts has increased twenty per cent. On such depression as this future prosperity may well be predicated. In 1880 the average price of anthracite foundry pig-iron, in Philadelphia, was $28.50 per ton of 2,240 pounds; in 1883, $22.37, and it is now about $20. The prices of 1883 were eleven per cent. less than in 1880; the production was twenty per cent. greater.

Pig-iron assumes great importance as a *producing* interest, and it is often claimed that depression in this branch of industry is always accompanied by depression in all others; but this assumption is putting the cart before the horse. When depression in the iron industry is caused by any general check to the consumption of iron, it surely indicates wide-spread depression elsewhere; but when depression in particular iron districts is accompanied by such activity in others that the aggregate production increases, it merely indicates survival of the fittest—the substitution of new and well-placed furnaces for old and misplaced ones—lower cost of production, higher wages for more effective work; in short, an adjustment to new conditions corresponding to the process which is affecting railroads. As a producing interest the iron industry is of very slight relative importance. The whole force of men and boys, who were employed in the census year in the production of about 4,000,000 tons of iron, consisted of about 20,000 engaged in mining coal for the use of blast furnaces, 32,000 in mining iron ore, 42,000 in blast furnaces, and perhaps enough more in subsidiary employments to make up 100,000 in all. In the present year, 1884, railroad construction may not exceed 4,500 miles, against 11,591 in 1882, or a falling off of 7,000 miles, which represents a reduction in the demands for rails only of about 700,000 tons, and of iron for other railroad use about 300,000, or 1,000,000 tons in all; yet there is no falling off in the production of iron even approximating such figures, therefore the general consumption has vastly increased, while the railway consumption has decreased.

The next consideration upon which future prosperity may be predicated, sooner or later, is the demand which our increasing population must make on existing instrumentalities of production and distribution. Agriculture is now prosperous. The railway system is in process of adjustment to new and sounder conditions. Of manufactured goods there seems to be a moderate excess, but it is generally believed that if this stock were distributed in the usual way on the shelves of the dealers, and had not been permitted or forced to accumulate on the hands of the producers, it would bear no appearance of excess. It is the waiting for events, the question, "What next?" that has for a little time checked the customary circulation of goods, and has caused what was named, when it was predicted a year and a half since, a temporary "commercial paralysis." This paralysis has been finally caused by what a president of one of the soundest banks in New York has well named "a moral panic," to distinguish it from an ordinary commercial or financial panic. When will this paralysis end? No one can tell, but we may measure the demand which our present increase of population at the rate of nearly or quite 2,000,000 persons a year must make upon the existing instrumentalities of production and distribution, and perhaps we may then, at least, venture to guess when the whole procession of the trades will move on.

Let it be assumed that within a year, more or less, we shall have reached a state of equilibrium somewhat similar to that of 1880, when all existing railways were fairly well employed, all manufacturing establishments fairly well adjusted to the then existing demand, and all farmers of intelligence were prospering. Under such conditions, it of necessity ensues that for each child born one adult must seek a new place of shelter, and each immigrant family must be housed; for each family of five, one new cotton-spindle must be set in motion; a half a ton additional of iron must be made; thirty or forty additional pounds of wool must be converted into cloth; and all other branches of productive industry must be increased by the addition of new capital, $i.\,e.$, new machinery, new tools, and new appliances. At the same time, the railway mileage must be increased in the ratio of not less than 6,000 miles a year to serve the crossway traffic of the existing population and to open new fields for the increase.

This is the kind of constructive enterprise, having reference to increasing and to future needs, which is subject to great variations and to vastly greater fluctuations than the mere subsistence of an existing population. The work of subsistence must go on, and must always give constant employment to by far the greater part of the population. We are always within less than one year of starvation, and within two or three years of being naked. In supplying these daily wants, work must be constant ; but constructive enterprise may vary fifty per cent. at one period compared to another, and that lesser portion of the population which must be engaged in construction in any decade may be pressed to the utmost for six to seven years, and then be half out of work for three to four, during which period of cessation the enterprising ones betake themselves to new land. We may approximately measure this constructive force. We numbered 50,000,000 in 1880. The abnormal increase by immigration added to the natural increase gives us now 57,000,000, June 30, 1884. We are probably increasing now at the rate of 2,000,000 a year. Let it be assumed that a condition of equilibrium may be reached January 1, or by July 1, 1885, that railways are then adjusted, manufacturers fairly employed, and agriculture prospering, what construction will become necessary to establish the capital necessary for sheltering, clothing, furnishing with tools, and moving the products of 2,000,000 people? It will be observed that, whatever the measure of this demand may be, it will be wholly a new demand for labor. The bricks must be made, the timber must be cut, the ore must be mined and smelted, the people must be housed and furnished with machinery and tools, *before* they can even begin to sustain themselves and to produce for themselves the daily subsistence which they will require.

All existing capital being balanced to the need of an existing population, the first demand of additional population is for new capital to be saved and invested, all capital being a concrete form of labor saved for future use. We may, therefore, convert the capital required by 2,000,000 people by way of terms of money into men's labor; that is to say, if any are now idle how soon will constructive enterprise require all their work and a great deal more?

First, *Shelter.* Can the average demand of each new family of five persons for shelter be fixed at any less than a house or part of a house, costing $500, to each family, or $100 per person? The poorest New England factory tenement costs more than this, but in the South shelter costs less. If this is the lowest measure, the provision for the shelter of 2,000,000 people will cost $200,000,000. That is to say, this sum of money must be paid for the conversion of trees, clay, and ore into houses. The average earnings of all who are engaged in these branches of work are not less than $400 per year, at which rate this sum measures a demand for the work of 500,000 wood-cutters, brick-makers, metal-workers, artisans, and mechanics. If $500 per family of five persons is too much, the furnishing of the house may be included.

Second, *Railroads.* The next great provision to be made is for the construction of new railroads. This year we may reduce to 4,000 miles, but soon the average must go up to at least 6,000 miles. If 6,000 miles a year on the average are needed, they will cost not less than $25,000 per mile in hard money for construction and equipment, or a total of $150,000,000. The men who do this work are laborers, miners, metal-workers, and mechanics. A fair average of their earnings would be not less than $350 per year, but to be conservative we may use $400 as a divisor, and then this sum measures a demand for the work of 375,000 men. At five to the mile, which is now a fair average, 30,000 men will also be required to operate the new railroads after completion.

Third, *Clothing and Iron.* In order to supply 2,000,000 with cotton and woollen cloth, boots, shoes, and hats, and with 200 pounds iron per head, an expenditure in new factories and iron-works of not less than $30,000,000 will be needed, and at $400 this measures a demand for the work of 75,000 men, while 30,000 men, women, and children will be needed to operate the works after they are constructed.[1]

How can we measure the capital which will be needed in the 330 other branches of manufactures which have not been named, in order to begin to provide for the future subsistence of 2,000,000 people? It cannot be done with accuracy, but already we have measured a demand for the work of 1,000,000 of the existing population; and of this work, the provision for shelter, clothing, and iron, requiring the work of 600,000 persons, is absolutely new work in addition to any and all work now done. How many capable and competent workmen or workwomen are there now out of employment and seeking work anywhere? What was the greatest number in the most depressed period after 1873? Is not all the common talk of over-production the veriest nonsense, when within one or two years from any given date all that there is produced must be used in making preparation for increasing wants, while, on the other hand, skilled, competent, and sober laborers never lack employment? This country cannot stop. The greater the check to constructive enterprise *now*, the greater the activity must be in one, two, or three years. Who shall say when it will begin? You may ask the apple-woman. Such are the facts of

---

[1] For the average earnings of the classes named, reference may be had to the census.

the past and present. Who dare forecast the immediate future?

How much will the progress of this country be hindered by distrust of politicians who are not statesmen, and by the futile attempts of ignorant legislators to regulate railway traffic and to control great industrial forces by means of meddlesome statutes inconsistent in their own provisions, and retarding rather than promoting the welfare of the people? Who can measure the iniquity of the railway wreckers who have used their power for nefarious purposes —stolen franchises, cheated their stockholders, and perverted the most powerful and beneficent instrumentality ever placed at the disposal of man to the basest purposes; and who have made it possible to say that the present is "a moral panic," caused by the want of honor and integrity among those who had secured places of highest trust and responsibility? The riots of Cincinnati had their origin in the perversion of justice in the criminal courts. The injury which has been inflicted by the perversion of the civil courts of a neighboring State, at the instance of men who can now be named, may some time culminate in a remedy equally disastrous, but by which the wrong will be remedied. When courts and judges are corrupted, men fall back upon their natural rights, and remedy their wrongs by rougher methods than those which are contemplated by law.

It would, of course, have been impossible for me to have given these facts had not the railway problem possessed a certain fascination which has lately led me to continue these tables, which attracted a good deal of attention about four years since, down to the present time. These tables are now placed at your disposal,[1] and I think they fully sus-

---

[1] This treatise was first prepared as a continuation of the testimony given by the writer before the Senate Committee on Labor, to be included in their final report.

tain the position which I have taken, to wit: that railway charges were reduced to the lowest possible terms about the year 1880, since which date they have sometimes been forced below the cost of the service; and that when the whole traffic of two great States like New York and Ohio is performed at a profit of a quarter of a cent a ton per mile, and sometimes for much less, the only effect of legislative interference would be fraught with danger, unless limited to securing publicity of accounts and a board of friendly arbitration, like the railway commission of Massachusetts, whose example in this matter has been followed in several other States.

In conclusion, a few words may be added upon the general principles upon which progress in the past has been based, and upon which it may be predicated in the future. What effect have these and other changes had upon the mass of the population who labor for wages, and whose daily bread depends upon their daily work? All profits, wages, and taxes are and must be derived from the conversion and sale of the annual product, *i. e.*, from the products of each succession of the four seasons. The money measure of this product—that is to say, its market value, is determined directly or indirectly by its competition with other like products in the great markets of the world. The world subsists by the exchange of product for product, and the balances are settled in money in the centres of exchange, whether national or international. The rates of wages are, therefore, a result corresponding with and measuring the share which the workman receives from the sale or exchange of the product on which he has spent his work.

In this country, the most effective machinery and the most versatile and intelligent labor are applied to the most ample natural resources possessed by any nation. A huge abundance, therefore, ensues from the least amount of human labor. On some of the fattest land in the West the measure of the product of one man working the best machinery with a pair of horses has reached *one hundred tons* of corn in a single season. The aim of some of the great "bonanza wheat farmers" of Dakota has been to apply machinery so effectively that the cultivation of one full section, or 640 acres, shall represent one year's work of only one man. This has not yet been reached, but so far as the production of the grain of wheat is concerned one man's work will now give 1,000 persons enough for a barrel of flour a year, which is the average ration. One man's work for one year suffices to supply 500 people with the largest quantity of iron consumed by any nation on the earth. The work of one operative in a wollen or cotton factory, or in the auxiliary print-works or bleachery, suffices to convert cotton and wool into textile fabrics for 250 or more of the most amply clothed people in the world. In proportion to the increase and efficiency of capital, a less number of laborers suffices for necessary work, wages increase in amount, and the purchasing power of each dollar of the earnings of the people is also augmented.

The truth of the fundamental law of labor is historically sustained—to wit: that high rates of wages of the highest producing power are the necessary result or correlative of the low labor cost of production. The operative of to-day earns twice the wages in ten hours that the operative of forty years ago could earn in thirteen hours per day. By the combined force of more adequate capital working harmoniously with more intelligent labor, the standard of a good subsistence is raised, the cost is decreased, the hours of labor are shortened, and the struggle for life is rendered less and less severe. The time will surely come when, by the working of these great forces of industry, intelligence, and integrity, even the vision of the mistaken enthusiast, who seeks to shorten the hours of work by meddlesome statutes, will be realized, and when eight hours of average work per day may suffice to produce an ample subsistence. In the meantime, the demagogue and the quack will mislead honest but mistaken men, and will endeavor to secure position and power by cheating them with paper money, crippling the railway service by means of so-called "anti-monopoly" statutes, and by restricting the freedom of contract by meddlesome statutes affecting the hours of labor of adult men and women, which retard more than they promote the object for which they are enacted.

It will be observed that the increase of population in this country is subject to a moderate variation from a uniform rate, as the immigration may be large or small; and that, conversely, immigration is retarded or stimulated by the conditions of industry and of constructive work. On the other hand, while the aggregate increase in railway mileage has been vastly greater than the increase in population since 1865, it has also been subject to very much greater fluctuations. On the 1st of January, 1865, the population probably numbered 34,000,000. On the 1st of January, 1885, it will probably number 58,000,000. The railway mileage on the 1st of January, 1865, was 33,908. Estimating the probable construction of the present year, 1884, at not over 4,000 miles (possibly 5,000), the number of miles January 1, 1885, will be a little over 125,000.

The following diagrams will show the great fluctuations, or *waves*, as they may be called, in the construction of railroads and in the consequent employment of labor.

Now the construction of a railroad represents in greater measure than almost any other form of capital a given and

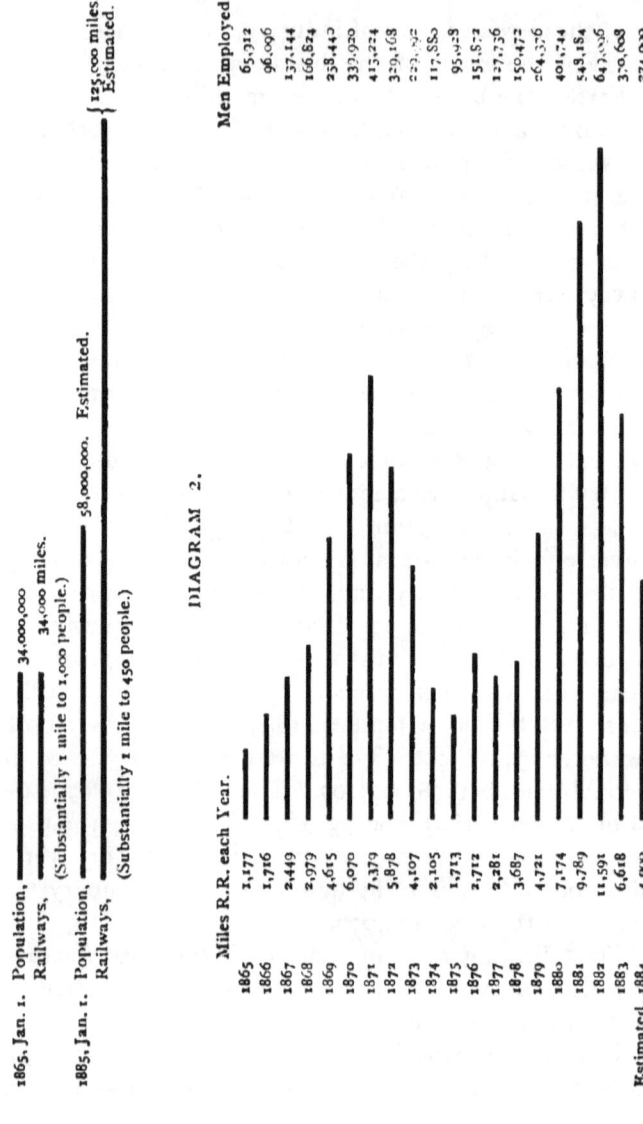

measurable amount of direct human or manual labor, coupled with but a moderate application of labor-saving machinery or capital. It represents more than almost any thing else a conversion of human labor, but little assisted by capital in labor-saving machinery, into one of the most effective forms of fixed capital. It is the work of the digger and the delver, of the navvy, the track-layer, and the woodman who cuts the ties, as well as of the iron- and coal-miner, the smelter, and the operative in the rolling mill, supplemented by the work of a relatively small number of mechanics in building stations and equipment. We can only reduce the construction of a railway to terms of so many men's labor for a given period in a very broad and general way, but even in this manner we may make an approximate estimate of the force employed one year on each mile. If we assume that, without paying any regard to the nominal amount of security issued, each average mile of railway construction has cost $25,000 in gold, this sum represents, or might be converted into, 50 men for one year at $500 each, or of 62.6 men at $400 each. A fairly approximate measure of the number employed would be midway, or 56 men. If the average pay is less, the number of men will be greater per mile per year. At this high ratio of wages, the force employed in the construction of railways has varied in the proportions set against the mileage table in the preceding diagrams. I use intentionally, in this case, a high average rate of earnings and probably a low money cost per mile, in order not to exaggerate the number of men employed.

Nothing is claimed for this computation except that it gives an approximate indication of the fluctuation in the demand for common labor which ensues from the activity or the increase in railway construction. It may be admitted that some labor-saving contrivances have been introduced in this branch of work since 1865, and that the construction of 1883-4 would represent a less proportionate number of men per mile at higher wages than in 1865, when the upward wave began to move, or in 1871, when the first wave began to recede; yet after making all due allowance for this variable term, railway construction remains in great measure an example of arduous manual labor in grading the way, piercing the tunnels, levelling the hills, cutting the ties, and in mining the ores and coal for the making of the rails. It is almost wholly direct human labor. The variation in this demand for labor cannot have been less than from about 60,000 men in 1865, up to 400,000 in 1871, down to 90,000 in 1875, up to 650,000 in 1883, and back to not over 280,000 in 1884, even if we exceed the estimate in the table of 4,000 miles and actually reach 5,000 miles in 1884. If we build only 4,000 then the demand will fall to 224,000. In this example we have an extreme case of the dependence of the common laborer upon the continuation of constructive enterprise; using the term constructive as a designation of that part of the work of the country which is quite distinct and separate from the necessary work of providing or moving subsistence for a given or fixed population at a given time. The subsistence of a fixed population, and the maintenance of existing capital or instrumentality of production for a fixed number of persons, constitute *necessary work*, which cannot vary or fluctuate in any great measure, whether "the times," so called, are "easy" or "hard." The great fluctuation in the demand for labor occurs in the demand for that small part of any given population which is or should be customarily employed in those constructive enterprises which are undertaken either for the purpose of meeting the increasing demands or "progressive desires" of an existing population, or in preparing to meet the absolutely necessary wants of an increase of the population.

Under the ordinary or normal demand of a time of long-continued peace, it may be safely assumed that at least ninety per cent. of all the people of this country, who are engaged in production or distribution at any given time, are in fact employed in providing or distributing the necessary means of subsistence, or in repairing or maintaining existing capital, or in keeping up the condition of farms to the standard of that time; and that not exceeding ten per cent. are or can be employed in adding to the capital or wealth of the country, or in making preparation for housing and furnishing the next year's increase of people and getting them ready to become themselves self-sustaining. According to the census of 1880, the total number of persons who were occupied in the production or distribution of the annual product, as well as in the constructive enterprises, was 17,392,099.

In agriculture	7,670,493
In professional or personal service	4,074,238
In trade and transportation	1,810,256
In manufacturing, mechanical, and mining industry	3,837,112
	17,392,099

14,744,942 males, 2,647,157 females.

If the proportions which I have adopted are approximately correct—to wit: ninety per cent. engaged in providing necessary subsistence and maintaining existing capital; ten per cent. engaged in constructive enterprise to meet increasing wants and an increase of population, then the proportion of the whole number in each department of industry would be:

Class 1—Necessary subsistence	15,652,899
Class 2—Constructive work	1,739,200

But it will be observed that hard times increase the number engaged in agriculture by giving some kind of occupation to those who are thrown out of other employment; we must, therefore, treat only the other occupations, numbering 9,721,606; or we may assume at a year's later date, when the "boom" of 1880 ended and the present depression began, the number was 10,000,000 persons occupied otherwise than in agriculture. Of this number we may also assume that the work of 9,000,000 could not cease and has not ceased even in the worst period of 1883 and 1884, because their work is necessary to mere existence; but the work of 1,000,000 engaged in constructive enterprises, depended wholly upon *the confidence of the owners or capitalists in the future progress of this country*. It is this point which I wish to bring into the clearest light. There has been and can be no lack of capital. In 1873 and 1883 the silly cry of over-production has been heard in the land. Over-production is but another name for an excess of capital. The times are "hard" or "easy," and prosperity or adversity depends on the single question whether constructive enterprise, or the preparation for future wants, is giving employment to the excess of capital and to one million persons, or to only half a million; half a million constitute only one per cent. of the population of 1880. It is the Micawber example on a grand scale—a shilling over is wealth, a shilling under is poverty. One per cent. of the population out of work (500,000) is adversity; one per cent. more workmen needed but not readily found is prosperity. Activity in the circulation of capital and labor rather than mere accumulation, indicates welfare; lack of confidence, slow movement, hard times, mean want because a fraction are unemployed, but that fraction is an army 500,000 strong.

I have already proved that there were about 300,000 less

laborers employed in the construction of railroads in 1875 than in 1871, and that there are probably over 400,000 less in that work in 1884 than there were in 1882.

In a previous part of this treatise I have shown the smaller relation which the capital in all manufactures bore to the capital in all railroads in 1880. We may assume that the cessation of constructive enterprise — that is, in the building of new factories, mills, and works of all kinds—was checked after the panic of 1873, and we know that it is checked now in 1884, but perhaps not in as great a measure as the construction of railways has been checked. In this cessation of factory building we may perhaps account for the lack of employment of a less number than those who have been discharged from railway construction, but probably enough to carry that number which I computed at 400,000 up to 560,000 in all, corresponding substantially to one half the total number of our present population which I have assigned to constructive work under normal conditions. Upon this very apparent condition of adversity we may predicate a speedy return of activity and prosperity. *Where can one find* 560,000 *men and women out of employment at the present moment?* They are not to be found, nor any number approaching such a maximum. Neither could any such number have been found in the darkest period of depression after the panic of 1873. The thing which really happens after one of these checks to the construction of railways, factories, works, or furnaces, is that those who are discharged from this class of work betake themselves to new land, open new farms, build up new towns in far-away places, and presently add to the demand for new railways and become consumers of metals and fabrics for which, again, more new works and new mills must be provided, even in addition to those required for the increase of population. Witness Western land sales in hard times.

It thus happens that the more severe the shock to constructive enterprise and the greater the depression at one period, the greater must be the activity and progress a little later. Thus it has been in the past, thus it will be in the immediate future. The one point which cannot be determined is the exact date when this change will come. This is a mental and not a material question—a question of confidence and not of capital. It is in the interval of adjustment to changed conditions that trade is dull and that "times are hard." The imagination is one of the most potent factors in rendering adverse conditions more intense, and in pushing favorable conditions to a dangerous extreme.

I have endeavored to show how important a factor the railroad is in all the work of modern life in this land. There is one more comparison yet to be made. The figures in the admirable census volume on transportation practically concur with those of Poor's Manual, each sustaining the substantial accuracy of the other. According to the census the number of miles of railroad in operation June 30, 1880, was 87,801, and the number of men employed in their operation and maintenance was 418,957, a fraction less than five men per mile. I have computed the number engaged in the construction of railroads in 1880 at over 400,000 men, which is apparently correct. If this estimate be accepted, more than 800,000 men were engaged in the railway service in the census year, and in the year 1882 the number must have been over 1,000,000. The total number of males engaged in all occupations listed in the census was 14,744,942. It therefore follows that one man out of every 18½ men occupied in any kind of work in this country, either mental or manual, was employed in 1880 in connection with railroads, and since then the proportion has been greater. Not less than 600,000 are now employed in the operation of railroads.

For many years more than one man in every ten men employed in any kind of gainful occupation, aside from agriculture, has been engaged either in constructing or operating railways. The importance of this fact can only be comprehended by comparing this great peaceful force, which is continually engaged in making the struggle for life easier, with the occupation of as great a number in other countries. Our army is but a border police force, opening the way for yet more abundant production. As the railway has diminished the cost of moving our great crops, our market has extended; every dollar thus added in money or in money's worth has been so much added to the annual product from which all profits, all wages, and all taxes are alike derived.

On the other hand, with every year the nations upon the continent of Europe have been more and more oppressed by increasing armies, heavier taxes, and increasing debts; each short interval of peace barely suffices to enable great armies to be recruited, but neither during peace nor war can debt or taxes be reduced.

The sum of our taxes—national, State and municipal—is eight per cent. upon the largest estimate of our national product ; but from the worst tax of all, the blood tax of a standing army, we are saved. We spend our force in building railroads instead of wasting it in passive war. In the principal states on the continent of Europe one man in every twenty-two is a soldier in active service in a standing army, and perhaps one more in every twenty-two is engaged in sustaining that soldier. The relative burden of the standing army is pictured by these two lines :

Europe,　　　1 in 22 ━━━━━━━━━━━━━━━━━━━━━━━━
United States, 1 in 400 ━━ in 1880, and now a less proportion.

These lines may well be pondered by those who treat the

rate of wages. It is the sum of wages by comparison of which the cost of production may be measured in money and not by the rate. When the obstructions of time and distance are removed, as they are almost wholly by railroads, the *rate* of wages will be highest in money where the cost is lowest in labor; because at that place the greatest skill, the best machinery, and the most productive natural conditions will be made use of, whereby the largest production will be assured at the least cost of human labor; when this product comes into competition with other products of like kind, the price will be the same provided the quality is equal, or higher if it is better. Witness the competition in London of the wheat of Dakota with the wheat of Russia and India. The measure of the cost is not the high rate of the wages of the few skilful men who work the machinery of production in Dakota, nor is it the low rate of wages of the peasants of Russia or of India. Wages are but the laborers' share of the value of the joint product of capital and labor, converted into terms of money by the sale of such product; the competition of capital with capital tends to a constant increase of product, coupled with a decrease in the rate of profit, while conversely the share of the laborer tends constantly to an increase both absolutely and relatively. Hence the more the capitalist applies his capital and increases his wealth the more the laborers' wages rise in rate and in purchasing power alike. The measure of the division of the laborers' constantly increasing share among themselves—the personal rate of each man's wages—rests wholly on the individual skill and industry of each member of the great industrial army. The man or woman who applies machinery most effectually, and who compasses the largest product with the least expenditure of time or labor, earns the highest wages,—in other words, obtains proportionately a larger share of the proceeds of the work than the one who is incapable or who is subjected to more adverse conditions.

Now what is the connection of these lines picturing standing armies with either the sum or the rate of wages? Each year's product is exchanged or sold and thus converted into terms of money; a small portion of the last season's crop, or money, is brought over to set this year's crop in motion, and a small portion of this year's product or avails in money is carried forward to the next; subject to these conditions each year's work must meet each year's wants, and the world is always within one year of starvation, the most prosperous nation within two years. Each year's product is converted into terms of money by sale or exchange, and from this sum must be derived all profits—all wages and all taxes. The sum of the product will depend upon the measure of labor which is applied to natural resources; if one man in twenty is withdrawn from productive work, by so much is the product decreased; if one other man's product is needed to sustain the idle soldier, by so much are the taxes increased. Wages are cut down in both ways, by reduction of the product and by the waste of what is produced in productive taxation 'or preparation for war. When the writer first compiled the article upon the Railroad and the Farmer, of which this treatise is a continuation, he submitted the following table.

Since that date (1881), with the possible exception of Italy and Holland, all nations upon the continent of Europe have either been subjected to a heavy increase of taxation, or else to an annual deficit and an increase of debt. It appears to be as impossible to sustain the present burdens of passive war as it is to disband the armies without revolution; yet migration is obstructed in order that the ranks may be kept full.

THE BURDENS UPON EUROPE AND AMERICA COMPARED (OMITTING RUSSIA, TURKEY, AND ALASKA.)

Relative Areas.

Europe, omitting Russia and Turkey, 1,546,802 sq. miles . . . .
United States, omitting Alaska, 3,034,399 square mile . . .

Relative Population to One Square Mile.

Europe, 145 per sq. mile.
U. S., 16½ "

Relative Burden of Debts to Each Inhabitant.

Since 1848 the debt of Europe has nearly trebled and is still increasing. In 1880 it was $16,794,800,000, or an average to each inhabitant of $74.64. Since 1880 it has increased, ...
In 1848 the United States owed no debt of any moment. On the 1st August, 1866, our war debt was at its maximum, and was estimated (liquidated and unliquidated) by Secretary McCulloch at $2,997,386,203 —an average to each inhabitant at that date of $83.35 . . . .
March 1, 1881, the debt had been more than one third paid, and was reduced to $1,879,956,412—an average to each person of $36.85 .
At this date the debt has been reduced to $1,450,000,000—an average to each person of $25 . . .

What do these lines mean to him who can read what is written between them? Is there not, on the one side, passive war alternating with active war, heavy cost of production, high taxes, low wages, misery and wrong, culminating in socialism, communism, nihilism, revolution, and repudiation; on the other, peace, order, and abundance, low cost of production, high wages, ample profits, stability, and welfare? *But this is on one condition*—that our intelligence is equal to our opportunity, and that the demagogue and the ignorant sentimentalist do not combine to tamper with the standard of value and debase our coinage; that the great forces of capital and labor are not prevented from working in harmony by meddlesome statutes; and that all taxes which the people pay are received by the Government and are honestly expended in the public service, by officials chosen and maintained in office on the condition only that their ability and character entitle them to serve in public offices. Let us now return to our main subject, the influence of the railroad.

No more facts or figures are needed to prove how profoundly a "moral panic" in railways must affect all interests in this country, and how much will be gained in human welfare if the railway service is now brought to the same standard of commercial integrity as that which controls all other enterprises in this and other civilized countries.

In the construction and operation of railroads the greatest ability, industry and, integrity have been and will be exercised; but it has been truly said, "the integrity of the many makes the opportunity for the fraud of the few." In railway enterprises the opportunity is greater in proportion to the complexity of the work and the magnitude of the sums employed, therefore have the villany, the fraud, and the breach of trust been almost measureless. Among even those who are now engaged in this work, every one who is in any way conversant with affairs can designate men whose names are synonyms for all that is able, honorable, and true. But alas! other names may also be given which are synonyms for all that is criminal, base, dishonorable, and fraudulent—names of men with whom no other man can serve without being himself defiled. The "moral panic" will end only when such men are not only dishonored but discredited. At that date confidence will be restored and constructive railroad enterprise will once more begin.

In another part of this treatise I have given an analysis of a loaf of bread, and also some facts in regard to the quantity of human labor represented by the wheat of which the flour is made. I proved that one man's work for one year, on a great Dakota farm, corresponded to the wheat required to produce 1,000 barrels of flour, and to deliver it at the railroad with an ample supply retained for seed. Let us follow this matter to the end: 4,500 bushels of wheat hauled from far Dakota to Minneapolis, there converted into 1,000 barrels of flour, and thence hauled to New York, is equal to an average haul of 120 tons about 1,700 miles. Upon the New York Central & Hudson River Railroad, the average number of tons hauled per year, per man employed in the freight service, is almost exactly equivalent to the work of one man for one year in hauling 120 tons 1,700 miles. I have not the exact data of the labor, in days' or years' work, in milling and preparing barrels, but as nearly as I can compute it this again is in the ratio of one year's work of one man to each thousand barrels. Add to the work of these three all the labor required to keep the machinery of the farm, of the flour mill, and of the railroad in repair, and the work of delivering the flour to the baker in New York, and even then we have not exceeded four years' work of four men to each thousand barrels of flour ready at the oven for conversion into bread. I have given the name of Samuel Howe, who sells good bread at a fair profit, and yet at a price of three cents and a fraction per pound, and from him I learn that only three persons are needed in his bakery and in his shops to convert into bread 1,000 barrels of flour and sell the same.

This, then, is the modern miracle: that by means of capital in farms, flour mills, railways, and bakeries, seven men, earning for themselves a good subsistence, serve one thousand persons with all the bread they need in a year, and, in the whole progress, from the planting of the seed until the bread is taken from the oven to be moved to the shop for sale, not one human hand will have touched the wheat, either in the grain or in the flour—only the bread itself will be handled. Yet not only the railway corporation, but the great farmers of the far West and the owners of the wheat elevators and of the flour mills, and, I dare say, the great baker of New York, have been the special mark for the obloquy, abuse, and interference of the demagogue, the sentimentalist, and the ignorant and meddlesome legislator; while capital is charged with oppressing labor and grinding the faces of the poor. It was said of old time that "the fool shall be brayed in a mortar." Perhaps the true punishment of those who excite passion and prejudice against these great forces of capital, by which bread has been made abundant and cheap, would be to deprive them of their benefit, and to force them to bray their own wheat in a mortar in order to gain their bread.

When the time of the National Legislature is taken up by the discussion of yet more obnoxious measures of national interference and futile attempts to control this great work, legislators may well remember that by means of the publicity of accounts which has been secured by the railroad commissioners of several States, and the yet greater national publicity of accounts secured by the private publication of Poor's Railway Manual, the service of the railways has been analyzed and defended, if this presentation of facts constitutes a defence and a justification of their great work. A commission which may bring public opinion to bear upon railway corporations may well be established, and there the work of the legislator may well cease.

There is another popular prejudice in refutation of which a few words may be said, to wit: the prejudice against the grants of great areas of public land to railway corporations. That this system has been abused may not be denied; that it has led to many premature schemes and to bad methods of construction by speculative construction companies is admitted; but this does not touch the merit of the system itself. That merit is this: by granting only each alternate section of 640 acres for a certain distance on each side of the line of construction, the subdivision of land in moderate parcels has been assured, and a monopoly of land has been prevented in a more effective manner than could have been compassed in any other way. It may be that some land grants ought to be forfeited for cause, and it may be that this grand ruling idea of the system has been sometimes evaded. Upon these mere incidents the present writer has nothing to say. He would only call attention to the fact that the system has worked well in causing a wide distribution of our population, and that it has assured a homestead to a vast number of persons who never could have attained one by any other method; because without the railroad, the construction of which has been induced by the land grant, the settlement of the land itself could not have been made. On the other hand the Government itself has gained the benefit of innumerable sales of the alternate sections at double the prices of its other unoccupied territory. It had been my intention to append a table to this treatise, giving the important facts in respect to the sales of railway and Government land on the lines of Land Grant roads, but

I found it impossible to find the data required. The subject might well be investigated officially by the Department of the Interior.

I have ventured in this treatise to give the reasons why we may expect a speedy return of constructive enterprise, of active employment, and of the quick circulation and rapid consumption of commodities, in which prosperity consists. It is doubtless true that "there is a tide in the affairs of men, which, taken at the flood, leads on to fortune," but it is also true that he who attempts to forecast the exact time when the ebb shall cease and the flood shall come, and who makes an error even of a single month, may lead to the loss of the fortune already gained, because no man can tell when a "moral panic" will end, or when confidence will be restored on which the whole depends.

APPENDIX I.

IN this connection the following analyses of the items which go to make up the price of bread in Boston may not be without interest. It will be observed that I have made use of the elements of cost in a small bakery, where the proportion of labor, fuel, etc., is much greater than in a large and thoroughly equipped establishment. I have also given the prices which are charged for a poor quality of bread in small shops in the poorest districts of the city. The destruction of the very poor is their poverty and their consequent inability to buy their food and fuel on good terms. What we greatly need in Boston is the counterpart of the "Howe National Bakery" of New York. At their great shops, which have been placed in three or four of the most densely populated districts of New York, a loaf of the best quality of bread, weighing two pounds before it is baked, and about $1\frac{3}{4}$ pounds afterward, is sold over the counter for cash at six cents per loaf, and at this price the owner of the bakeries is satisfied with his profit. In his works the cost of labor and fuel is less than half the sum pictured in the diagram which gives the cost of bread to the poor of Boston. Again, in this we find an example of adequate capital—high wages to the operative in the bakery, low cost of production of baked bread, and cheap food to the poor under the law of unrestricted competition, and under the rule of service for service, by means of which society itself exists, and under which labor and capital work as allies, not enemies.

The following analysis was submitted by the writer to the Committee on Education and Labor of the United States Senate:

Analysis of Cost of a Loaf of Bread.—I am prepared to admit that the railway has been a most important factor in distributing food among the people of this and other lands, for without it thousands might starve, but I shall also prove to you, in the analysis of the loaf of bread, that it has become relatively the factor of least importance, at its present cost, of all the items which constitute the cost of bread to the consumer; therefore, before you undertake to regulate the railways and thereby to reduce the price of bread, meat, and fuel, you must give your attention to vastly greater elements in their cost, which may be more readily made subject of statute law than the railway service can be, if either kind of work is to be taken in charge by the State.

I shall take as my unit 450 bushels of wheat to be converted into 100 barrels of flour and then into bread, and I shall present to you all the elements of the cost of this bread, both in figures and by graphical illustration, as follows:

What makes the price of bread in Boston? Four hundred and fifty bushels of wheat are required to make 100 barrels of flour. In the left-hand column it is assumed that this wheat has been raised near Chariton, Iowa, and milled in Chicago. In the right-hand column it is assumed that the wheat has been raised near Glyndon, in Dakota, and milled in Minneapolis.

It will be observed that if the railways earn as profit 30 per cent. of their charge, their profit on each barrel of Iowa flour moved about 1,500 miles is only $35\frac{1}{4}$ cents, and on each barrel of Dakota flour moved nearly 2,000 miles, only $59\frac{1}{4}$ cents. In point of fact the actual profit on grain and flour carried long distances is much less than 30 per cent. of the charge, and the actual profits for the above distances does not probably exceed 25 cents per barrel and 50 cents per barrel, respectively.

The railway charges are now so small that it does not leave you much of a margin to work upon and to save, but you cannot fail to notice that the charges made by the bakers and grocers is very large, and gives you an ample margin for legislative action. If you reply that all attempts to regulate the price of bread have failed, may I be permitted to rejoin that all attempts to regulate the charge of the railways have also failed, except, perhaps, in

Chariton, Iowa.   October, 1883.   Glyndon.

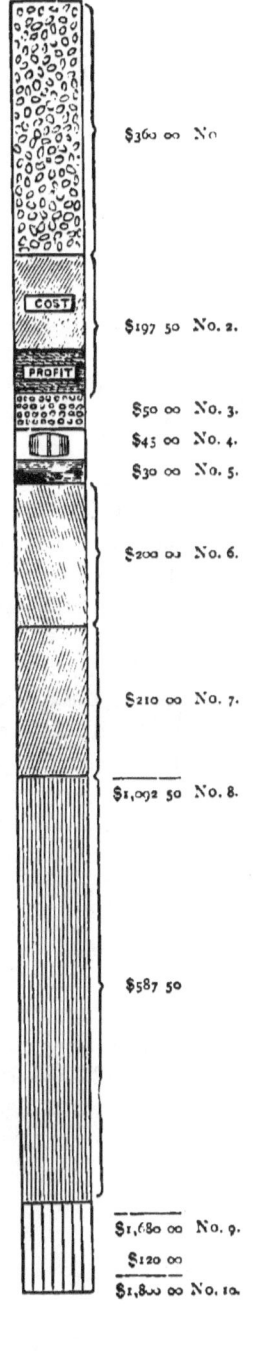

No. 1. $405 00   No. 1, $405, is the price which the farmer receives in Iowa, at 90 cents per bushel; $360, in Dakota, at 80 cents per bushel. $360 00   No. 1.

No. 2. $117 50   No. 2, $117.50 is the charge made by the railway for moving 450 bushels of wheat from Chariton to Chicago, and 100 barrels of flour thence to Boston, $197.50; Glyndon to Minneapolis and thence to Boston, $82.25; cost of railroad service at 70 per cent., $138.25 of the total charges. $197 50   No. 2.

$35.25 profit, at 30 per cent., $59.25.

No. 3. $50 00   No. 3, $50, cost of milling. $50 00   No. 3.
No. 4. $45 00   No. 4, $45, cost of barrels. $45 00   No. 4.
No. 5. $30 00   No. 5, $30, merchant's commissions and cartage in Boston. $30 00   No. 5.

No. 6. $200 00   No. 6, $200, cost of labor in making 100 barrels flour into bread in a small bakery. $200 00   No. 6.

No. 7. $210 00   No. 7, $210, cost of fuel, yeast, salt, etc., used in converting 100 barrels flour into bread. $210 00   No. 7.

No. 8. $1,057 50   No. 8, final cost of bread ready for distribution, average 3½ cents per pound; varying a little with the quality of the flour and the quality of bread. Iowa flour yields 270 and 290 pounds per barrel; Dakota flour yields 280 and 300 pounds per barrel. $1,092 50   No. 8.

$562 50   No. 9, the price which the poorer people of Boston pay for poor bread, made from a medium grade known as "baker's flour," averages not less than 6 cents per pound, which makes the cost of distributing 100 barrels of Iowa flour baked into bread, No. 9, $562.50, and 100 barrels Dakota flour, $587.50 at the minimum yield of 270 and 280 pounds bread to the barrel. When either kind of flour is treated so as to yield 300 pounds bread to a barrel and sold at 6 cents per pound, $80, or $120, is added, and the final cost of the bread to the consumer is at the rate of $18 per barrel of flour, No. 10. $587 50

No. 9. $1,620 00 $1,680 00   No. 9.
         $80 00 $120 00

No. 10. $1,700 00 $1,800 00   No. 10.

Belgium, where the Government has once at least been obliged to prohibit the private corporations which own a part of the railroads from *lowering* their charges, lest the Government railroads should be unable to compete with them.

Logical Consequence of the Demand for Governmental Regulation of Railroads.—Your committee has been asked, by what are known as the advocates of "anti-monopoly," to frame and present to Congress such laws as will forbid capital taking the advantage of labor by means of excessive charges for railway service, which charges are said "to make the rich richer and the poor poorer," and "to make bread dear."

The distribution of bread by bakers' wagons and through grocers' shops is, as I have said, simple but costly; the distribution of wheat and flour by railway is complex and difficult, but it is now done at so little cost as to leave little margin to be saved.

If your committee will first regulate the distribution of bread and reduce its price by statute, and, second, reduce the cost of barrels or require the substitution of cheaper sacks, you may then be fully prepared to frame suitable statutes for the regulation of the railway service. I recall this subject because the advantage of this method is that you can begin in Washington, and, by reducing the cost of living there, you can make the salaries of Senators and Representatives in Congress more adequate. When you have fixed the price of bread by legislation, you will, of course, take up meat, timber, and fuel, and after you have established an economic millennium in Washington, the several States, cities, and towns can supplement your national statutes by adequate municipal ordinances, in order to complete the system.

Effect of Railroad Charges on Cost of Meat.—I have not been able to make a complete analysis of the price of beef in Boston, but this much can be submitted. Texas steers, worth 4 cents per pound live weight at Emporia, Kans., can be and are brought to Boston at a charge of 1 cent per pound. What it costs to fatten and kill them I know not, but this I do know, that if the price of my sirloin is high the railway charge has little to do with its cost to me. Salt meat is brought at as low a charge as grain—hence railway charges have little to do with the high price of meat in the Eastern States.

It therefore follows that the monopolists, *if any such there are*, who are grinding the faces of the poor and rendering bread dear, are not the Vanderbilts, the Tom Scotts, or the Garretts, but they are the nameless bakers, grocers, and others, who have added this enormous charge to the cost of bread and meat. The whole railway service, from the field to the baker's oven, costs but half a cent per pound, but the service of the baker, and the grocer, and the shopman, costs $2\frac{1}{2}$ to 4 cents per pound of bread. If you will analyze your pound of beefsteak, or, if you are a Yankee, analyze the salt pork with which your beans were baked for your Sunday breakfast, I think you will find the greatest monopolists, *if any there are*, are running the butcher wagons and the provision shops of your cities. After you have succeeded in abating these enormous charges; after you have regulated the simple traffic of the baker, the grocer, the butcher, and the provision dealer; *after* you have prevented them from "grinding the faces of the poor," *then* take up the railway question, if you please, and see what is left for you to do. In dealing with this simple matter of the shopman and of the service of distribution by cart and wagon, you may learn how to regulate by statute the complex operations of the great railways of the United States, which have taxed the biggest brains and the ablest men of the land these twenty years or more. These men have laid the foundation upon which you can work.

APPENDIX II.

Upon one of the great farms of Dakota,

```
1 man in 1 day plows    4 acres at 20 bushels per acre   80 bushels.
   "       "    seeds  15    "      "     "     "    "  300    "
   "       "    harrows 15   "      "     "     "    "  300    "
   "       "    cuts   15    "      "     "     "    "  300    "
   "       "    shocks 10    "      "     "     "    "  200    "
   "       "    thrashes and draws to elevator,          40    "

300 bushels at the railway therefore stands for the work of 3¾ men plowing 1 day.
   "       "       "       "        "       "      "     "   1  man seeding  "
   "       "       "       "        "       "      "     "   1   "  harr'w'g "
   "       "       "       "        "       "      "     "   1   "  cutting  "
   "       "       "       "        "       "      "     "   1½ men shocking "
   "       "       "       "        "       "      "     "   7½  "  thrashing and
                                                                      drawing.
                              (Say 16 men.)  15¾ men 1 day, 300 bu.
```

Or, 18¾ bushels per man, at 20 bushels to an acre. Multiply by 300 working days in a year, and the equivalent is 5,625 bushels for one year's work of one man. Leave 1,125 bushels for seed and home consumption, and we have 4,500 bushels, from which 1,000 barrels of flour will be made. A year's annual ration of wheat flour to each person is one barrel a year, which will make 275 pounds of baked bread.

Let us assume 70 cents per bushel as the price of wheat at the railroad in Dakota, and produce 1,000 barrels flour into bread in New York. The various charges are as follows:

4,500 bushels wheat at 70 cents	$3,150
Moving to Minneapolis as wheat, and from there to New York as flour, at present rates (August, 1884)	1,440
Milling	500
Barrels	450
Conversion into bread—labor and material	1,750
Selling the bread over the counter	500
	$7,790

There may be a charge for a merchant's commission and for cartage if the flour is not bought by the baker from the miller, or is not delivered from the cars at the bakery	$210
275,000 pounds of bread, $2\frac{9}{100}$ cents per pound	$8,000

It will have been observed that the raising of the wheat represented the labor of 1 man for 1 year. The moving of the wheat and flour over 1,700 miles represents the direct labor upon the railway of $1\frac{1}{2}$ men working 1 year. The direct labor in milling and in making barrels from the log represents the labor of 1 man working 1 year. Add to the $3\frac{1}{2}$ thus far the work of 1 man 6 months, or $\frac{1}{2}$ man 1 year, engaged in the repairs of machinery, and 1,000 barrels of flour delivered in New York represent only the direct labor of only 4 men for 1 year.

In the Howe National Bakery of New York labor and material are economized to the utmost, and the bread is sold over the counter with the least waste of force. The conversion of 1,000 barrels of flour into bread and its sale represent the work of only 3 persons working 1 year. The modern miracle is that 7 men serve bread to 1,000 persons, and in so doing earn high wages for themselves, while the owner of the bakery earns his private fortune in selling good bread at 4 cents a pound or less. His six cent loaf which is upon the table before me weighs $1\frac{3}{4}$ pounds. None need ask better bread.

The entire profit of the railroads for moving 5,500 bushels of wheat 200 to 300 miles, and 1,000 barrels of flour 1,400 miles, at the present rates, has been computed by one of the most competent experts at $225—being $15\frac{1}{2}$ per cent. of the charge of $1,440. That is to say, the profit of the railway for bringing 1,000 barrels of flour 1,627 miles is $22\frac{1}{2}$ cents per barrel—just one half the cost of the barrel in which the flour is packed, or a trifle more than the value of the empty barrel in New York. In the diagram which I submitted to the Senate Committee, I gave the price of bread in the small shops where the poor deal in Boston—the cost of bread in a small bakery. I assigned 30 per cent. of the railroad charges to profits. Now I have exact data from the railroads, and

I use the figures of a baker who bakes at wholesale, but sells bread at the least charge. These figures prove that the poor of New York are served with better bread at much less price than are the poor of Boston.

Wheat assumed to be raised near Glyndon in Dakota: 5,625 bushels to one man's work ; 1,125 retained for seed or for domestic consumption.

	$3,150	Price of 4,500 bushels wheat at 70c., delivered at the railroad.
Total	1,215	Cost of railway service 4,500 bushels wheat 200 to 300 miles, 1,000 barrels flour 1,400 miles.
charge to N. Y.	225	Profit on the railway service.
	500	Milling 1,000 bushels flour.
	450	Barrels for 1,000 barrels flour.
	1,750	Labor, fuel, yeast, etc., used in making 275,000 pounds of bread.
	500	Cost of selling the bread.
	210	Incidentals.

$8,000 Cost of 275,000 pounds of bread in New York ; or, 7 men feed 1,000 for 1 year with bread.

If the labor of those who provide fuel and other materials for the railway and for the baker be added, the number might be raised to 10 men to 1,000 barrels of flour converted into bread.

At the risk of repetition let me again give other examples of the saving of labor which has resulted from the application of adequate capital and skilled labor. The year's work of 1 person is as follows : 1 in a cotton mill spins and weaves cotton cloth for 250 persons ; 1 in a woollen mill, woollen cloth for 300 persons ; 1 in a coal mine, iron mine, or iron furnace serves 200 pounds iron each to 500 persons ; 1 in a men's boot factory makes 2 pairs a year of boots or shoes for 800 persons ; 1 in a women's boot or shoe factory makes 3 pairs a year for 1,000 persons ; 1 in a shirt factory sews 2,400 excellent shirts, or more of lower quality, or 4 a year for 600 to 800 persons.

The poor sewing women are only those who sew in a poor way by hand. Skilful sewers in the shirt factory earn more than $10 per week. How much labor the materials used may represent is not included in these computations. In the case of the bread the wheat is traced from the beginning to the end. It may be admitted that this is an extreme case, and that the average production of wheat, except on these great bonanza farms, so-called, represents a much greater amount of labor. I have only presented the extreme of the present because it may become only the average of the future, but that part of the cost of the bread which constitutes the railway service leaves little margin to be gained until some new invention is applied by which the cost may be reduced. The profit of the railway on each pound of bread is $\frac{9}{1000}$ of a cent. It will puzzle legislators to cheapen this service; they may make it more costly.

The statement that the wheat from which 1,000 barrels of flour may be made, which represents the yearly ration of 1,000 persons, can be raised as the equivalent of one man's labor for one year, may be questioned. It seemed almost incredible to the writer until he had proved it by incontestible evidence of many competent witnesses. A fair *average* equivalent for one day's work of one man on a Dakota farm is $12\frac{1}{2}$ bushels of wheat in an ordinary season. On a well managed and thoroughly equipped farm in a season in which the crop is 20 bushels to the acre, the average for one day's work of one man has proved to be $18\frac{3}{4}$ bushels. This season, when the crop is expected to be 25 bushels per acre, it will be over 20 bushels per man per day. That is to say, the average per man per day is very nearly the product of one acre, whatever that may be according to the season. If we multiply the middle statement of $18\frac{3}{4}$ bushels per man per day by 300 working days, we have 5,625 bushels of wheat as the equivalent of the continuous work of one man for one year; but of course about three men will be employed for only part of a year, or during the wheat-growing and harvesting season. After the wheat farm has been fully equipped with adequate machinery and brought into good condition, the crop can be planted, made harvested, and moved to the elevator at a cost ranging from $6.00 to $10.00 per acre, according to relative conditions; it is claimed that on the best farms most completely equipped the whole cost can be covered at $5.00 per acre. It may be said that this cannot last, but such a hasty conclusion may not be warranted. There are as yet, no signs of exhaustion; the soil of this section appears to be of a peculiar kind. The frost strikes deep into the ground, and long before it is out below, the surface is dry, warm, and ready for the seed; after that the moisture from the melting frost keeps coming up laden with elements of fertility. How long this will continue who can tell? But even if it may only last a few years, then after that the division of the land into smaller farms will bring in fertilizers and other methods of economic cultivation. In the meantime what is the area available? The area of Dakota only is 150,000 square miles, of which but a mere fraction is yet under the plow, and north of it is the almost unlimited area of wheat land in Manitoba. Is it not apparent that wheat may go even below thirty-four shillings per quarter in Mark Lane before the supply of wheat from Dakota would cease to meet the demand, except the demand of our own country should stop the export tide? With our present railway and steamship service, even at paying or profitable rates of traffic, our farmers can unquestionably contest the markets of Europe with India and Russia, down to less than thirty-four shillings a quarter in Mark Lane, if they cannot do better at home. The English quarter of wheat by which prices are quoted is 480 pounds, or 8 bushels of 60 pounds each —thirty-four shillings per quarter will yield a little over one dollar per bushel in London, at which we can readily continue the traffic, but of course at a greatly reduced profit to the farmer. The India railways, for which a very large appropriation is about to be made, will doubtless render the competition in India a little sharper, but it will be observed that the system adopted has been planned mainly with reference to the distribution of food in India itself, for the purpose of preventing the recurrence of famine. It will therefore increase the consumption of food in India, and may diminish the export of grain to England instead of increasing it.

JULY, 1884.

OCCUPATIONS CLASSIFIED.[1]

There can be no better way of presenting the immense importance of this problem in respect to the distribution and use of food, while incidentally enforcing the need of manual as distinguished from purely mental instruction, than by classifying the whole force of persons who were engaged in gainful occupations in the census year according to the kind of work done by each class.

This force numbered 17,392,099, or a fraction less than one in three of the population. The list of their occupations is as accurate as the enumeration of the population itself, because it was made by the same enumerators. The only qualification to be made is that many laborers are listed as "laborers not specified," who may have been on farms; and doubtless many men are listed as mechanics, whose work was done in connection with a manufacturing establishment. In consequence of the latter fact, the separation of the mechanics or artizans whose work was "individual" from those who formed a part of a "collective" force employed in a factory, can only be made approximately.

The following table gives a very close approximation to the number of persons in each one thousand who were occupied in any kind of gainful occupation in the census year:

CLASS I.—Purely mental and individual work:
  Clergymen, lawyers, physicians, surgeons, chief officers of banks, telegraph companies, railroads, insurance companies, and other occupations of like kind 40
CLASS II.—Distributive; in part mental, in part manual, in part collective, in part individual:
  Merchants, tradesmen, clerks, hotel-keepers, commercial travellers, salesmen, and saleswomen 60

---

[1] These tables belong to and are to be considered in connection with the matter in Appendix VII., following the essay on wages. See p. 171.

CLASS III.—Manufacturing or mechanical of the collective order—that is to say, occupations in which large numbers of persons are concentrated in factories:
Textile factories, iron and steel works, machine shops, clothing, hat, boot and shoe factories, or other analogous works, 92 to 100, say . 100
CLASS IV.—Mechanical pursuits, mainly individual rather than collective :
Carpenters, blacksmiths, masons, wheelwrights, painters, etc., etc., 107 to 115, say 107
CLASS V.—Personal service :
Domestic servants, draymen, employés of railroad, telegraph, telephone, and express companies, steamboat men, sailors, waiters, etc. 131
CLASS VI.—Laborers :
Farm laborers (191), laborers not specified (107), miners (14) . . 312
CLASS VII.—Agriculturists :
Farmers, stock-raisers, etc. etc. 250
       Total . . . .[1] 1,000

It may be held that the food of the members of Class I., and the servants in Class V., will be intelligently purchased and used, and that a lessening proportion of those engaged in collective work, Class III., will be served with the economy of the collective or boarding-house system.

On the whole, it may be held that 900 out of each 1,000 will buy and use food according to the measure of their own personal faculty in the matter, and that the lower the grade of the workman, the greater the want of economy in buying and the greater the waste in use.

---

[1] If we adopt the classification of the census, we find the following proportions in each 1,000 of the people who are occupied in gainful occupations.

	Agriculture.	Professional and Personal.	Trade and Transportation.	Man'f'g, Mech'l, and Mining.	Total.
Southern States, inc. Delaware and Mo. .	646	196	63	95	1,000
Middle States, inc. N. Y., N. J., Pa. .	197	293	157	353	1,000
Western and N. Western and Territories	442	242	107	209	1,000
New England	192	223	134	451	1,000
Whole country	441	234	104	221	1,000

The absolute necessity for manual as well as mental instruction is proved in the most conclusive way by a comparison of the proportion which the work which is *individual* bears to the work which is *collective*. In the one case personal faculty, or "gumption," is the quality which assures success; in the other, long practice in a single one of many processes. In this again may be found the proof that the rate of wages is finally determined mainly by personal qualities, and rests at last on individual character, capacity, and moral integrity.

May not one find in the forces developed by modern science such an assurance of abundance that moderate intelligence, good health, and industry will certainly secure a good subsistence, in which case it may not pay to be rich?

The two conclusions which must be drawn from these tables are:

1st. The relatively small proportion of all persons engaged in productive work who have been able to reach a plane above that of the laborer or domestic servant, or of the small farmer who works harder for a meagre subsistence than any of his hired men.

2d. The small relief which has yet been given by the adoption of the collective factory system,—by the use of automatic machinery and by the division of labor.

In this country at least this relatively low plane on which more than one half the working people are still to be found, cannot be attributed to any lack of or monopoly of land. There is far more land waiting for laborers capable of gaining their subsistence from it than has yet been put to any productive use.

In fact, both land and capital are in such abundance that every person, capable either of using the land or of applying capital thereto, is being sought for by the representatives of railways and of other corporations.

Under such conditions can there be any thing wanting except those personal qualities which have been named on which the rate of wages finally depends, or by which the rate of wages is finally made,—character, capacity, and industry?

WHAT MAKES THE RATE OF WAGES?

(ADDENDA TO SECOND EDITION.)

Since the completion of the treatise upon "What Makes the Rate of Wages?" the attention of the writer has been called to the great importance of a correct analysis of the occupations of the people of this country, and to the necessity for such an analysis before any scientific treatment of the three great issues now before the public can become even possible.

These three questions are:

1. The Railway Service, and the proposed regulation thereof by the Government.
2. The Silver Coinage and the Acts of Legal Tender.
3. The Collection of the National Revenue.

All of these questions are but phases of the major issue in respect to the relations of labor and capital, or branches of the final question—What makes the rate of wages?

We may first consider the occupations of the people by sections, in their effect upon the traffic of railways, and for this purpose we may make use of the census classification.

The table of the occupations of all who were engaged in gainful employments may be accepted as one of the most accurate in the census of 1880, in view of the fact that the same enumerators who counted the population also made this list, and each person enumerated gave his or her own occupation.

The census classification is into four groups, viz.:

1. Farmers and farm laborers.
2. Professional and personal service.

3. Trade and transportation.
4. Manufactures, mechanics, and mining.

In the following table the relative proportions engaged in each occupation in each 1,000 are shown as to the whole country, and as to each section :

Now, if we consider this table *a priori*, what might we expect to find the relative railway traffic to be ?

In the New England States, where the manufacturing and the mechanic arts give employment to the largest number of persons, and where the population is dense, we should expect to find the largest number of passengers to each mile of railroad. In the Southern States, where population is widely scattered and is chiefly engaged in agriculture, and where almost all the crops are light in weight, we should expect to find the least number of passengers and tons of merchandise per mile.

In the Middle States, which are both manufacturing and commercial, and through which the heavy Western crops are moved, we should look for the greatest quantity of merchandise per mile, and in the grain-growing States of the West we should look for heavy traffic in merchandise and a small number of passengers per mile.

The facts fully justify the theory, and although the two following tables do not absolutely follow the same rule of sectional division, yet the analogy of the respective laws of distribution is very plain.

PROPORTIONAL MOVEMENT OF PASSENGERS AND MERCHANDISE SHOWN BY SECTIONS, THE DIVISIONS BEING MADE ACCORDING TO THE NATURE OF THE TRAFFIC.

SECTION 1.—New England States.—Food and Fuel, moved in ; Manufactures, moved out and distributed.

SECTION 2.—Middle States (New York, New Jersey, Pennsylvania, Delaware, and Maryland).—Food, moved in and through ; Fuel and Metal, moved out ; Manufactured Goods, moved out and distributed.

SECTION 3.—Western States (Ohio, Michigan, Indiana, Illinois, Wisconsin, Minnesota, Iowa, Kansas, Missouri, and Nebraska).—Food and

PROPORTION OF THE PERSONS IN EACH 1,000 OF ALL WHO WERE OCCUPIED IN ANY KIND OF GAINFUL EMPLOYMENT IN THE CENSUS YEAR. TOTAL NUMBER, 17,392,099. I. FARMERS AND FARM LABORERS. II. PROFESSIONAL AND PERSONAL SERVICE. III. TRADE AND TRANSPORTATION. IV. MANUFACTURING, MECHANICAL, AND MINING.

1. In the whole country.

| I.—F. & F. L. 441. | II.—Prof. & Pers. 234. | III.—T. & T. 104. | IV.—M. & M. 221. | 1,000 |

2. Southern (or lately Slave) States, including Delaware and Missouri.

| I.—F. & F. 646. | II.—P. & P. 196. | III. 63. | IV.—M.&M. 95. | 1,000 |

3. New England States.

| I.—F. & F. 192. | II.—T. & T. 134. | IV.—M. & M. 451. | 1,000 |

4. Middle States, exclusive of Delaware.

| I.—F. & F. 197. | II.—P. & P. 293. | III.—T. & T. 157. | IV.—M. & M. 353. | 1,000 |

5. Western and Northwestern States and Territories.

| I.—F. & F. 442. | II.—P. & P. 242. | III.—T. & T. 107. | IV.—M. & M. 209. | 1,000 |

Timber, moved out; Manufactures, moved in; Fuel, etc., distributed.

SECTION 4.—Southern States—Cotton, Wool, Hemp, Tobacco, and some Metal, moved out; Food and Manufactures, moved in; local distribution.

TABLE I.

	Miles R. R.	Passengers Carried.	Passengers per Mile.	Proportion per Mile.
Section 1.—N. E.	6,323	72,377,556	11,446	
Section 2.—Mid.	17,131	126,354,067	7,376	
Section 3.—West	60,525	83,823,759	1,385	
Section 4.—South	26,135	17,453,579	668	

FREIGHT MOVEMENT.

TABLE II.

	Miles R. R.	Tons Carried.	Tons per Mile.	Proportion per Mile.
Section 1.—N. E.	6,323	30,670,213	4,850	
Section 2.—Mid.	17,131	186,736,924	10,900	
Section 3.—West	60,525	144,853,216	2,393	
Section 4.—South	26,135	31,014,619	1,187	

TABLE III.

	Tons Carried One Mile.	Proportion to Each Mile.
Section 1.—N. E.	285,797	
Section 2.—Mid.	936,890	
Section 3.—West	356,585	
Section 4.—South	126,292	

The importance of this classification will be very apparent when we consider the relation which volume of traffic bears to rates of charge for such service.

While it does not follow absolutely that the more freight and passengers a railway system is called upon to move the lower may be its rates, yet it is a law that unless a railway system is worked up to the capacity of such equipment as it must have in order to work at all, its rates of charge must be higher in inverse proportion to the amount of its traffic. Hence it follows that after all due consideration has been given to grades, length of haul, terminals, fuel, and to the quality of the traffic ; and after the rates have become adjusted so as to meet all these complex and confusing elements of the problem, the number of tons and of passengers will then constitute a finally controlling element. Those railways which are in the great lines of movement of grain (now about 100,000,000 tons per year), fuel (now about 90,000,000 tons), timber, and other heavy substances, must and will be worked at a much less cost and at a much lower charge per ton than those railways whose traffic consists almost wholly of fibres (2,000,000 tons), metals (6,000,000 tons), and general merchandise or food for local distribution.

Under these conditions, the more the attempt is made to control the rates of traffic by statute, the higher the rates charged must be, because it is very plain that the rates on railways which have a small traffic cannot be reduced by statute to the level of those which have a heavy traffic unless the State takes them and operates them at a loss ; and therefore it follows of necessity that statute interference can only end in an advance of the low rates now charged on the lines having a heavy traffic to the higher rates of the lines having a small traffic, if the statutes do not prove to be inoperative. Such statutes have up to this time utterly failed, after a period of ineffectual disturbance to the whole traffic of the country.

The main interest, however, in the classification of occupations is the clue which it gives to the small compensation or rate of wages for which it appears that a very large portion of the people must work, because there is no more to be divided among them.

The classification of the census under four heads is of no value for this purpose, because it places domestic servants and common laborers in the same category with professional persons. We may, therefore, sort all persons occupied into seven groups :

For this classification into seven groups only approximate accuracy can be claimed, for reasons which are given hereafter ; but the main feature, to wit, the relatively low grade of the work, and therefore low wages of a very large portion of the people, is unpleasantly conspicuous. A little less than one in three of the whole population is engaged in gainful work, and if my computations are approximately correct the share which each person of each group of three of the so-called " working classes " can have is only what forty to forty-five cents a day will buy. When depression reduces even this low measure, what wonder that trouble ensues if injustice is even suspected in the social order by the uninformed or ignorant ?

In the following table the specific numbers in each separate occupation are sometimes given from the census data exactly, and sometimes by computing together in round approximate figures those whose occupations are analogous. The shading in the graphical lines is intended to show approximately the proportion of each class whose earnings may be above the average of annual income, as compared with those whose earnings are at that rate or below it. No absolute data exist for making this last separation, —it is by estimate only.

We may consider this table, not only because of the picture which it gives us of the planes into which society is now stratified, if such an expression may be used, but also in the relation of each class to the other in its purchasing or exchanging power ; and, finally, in the effect which a lack of occupation on the part of any large number in the lower planes must have upon the demand for the products of capital or upon the prosperity of those in the higher planes.

In Class I, consisting of persons whose work is purely mental, are to be found all teachers, country clergymen, literary persons, journalists and the like, comprising more than one half of the whole number ; and in this category will be found a very large

Class.	No. in each 1,000	Computed total No.	Approximate or Actual Numbers in Each Class.
1. Mental Work	40	696,000	Clergymen, 64,968; lawyers, 64,137; physicians and surgeons, 85,671; teachers and literary, 227,710; journalists, 12,308; engineers, 8,126; musicians, 30,477; officers of corporations, banks, railroads, insurance, and others, 202,468.
2. Mental and manual	60	1,044,000	Merchants and traders, 481,450; hotel keepers, 32,543; clerks, salesmen, commercial travellers, brokers, and all others engaged in the purchase and sale of goods, 521,898.
3. Automatic machinery	100	1,740,000	Collective factory work, textiles, printing and bleaching, 500,000; metals and machinery, 300,000; clothing, 450,000; boots, shoes, and hats, 210,000; all others, 280,000.
4. Hand and mach. tools	107	1,861,800	Mechanical, not collective, carpenters, coopers, and other workers in wood, 500,000; blacksmiths, 172,726; painters, 128,556; masons, 102,473; all others, 958,045.
5. Manual	131	2,279,400	Service, express, railroad, telegraph employés (not laborers), 300,000; domestic servants, 1,075,655; laundry, 122,000; waiters, 200,000; draymen, hackmen, etc., 180,000; all others, 391,345.
6. Horse and hand tools	250	4,350,000	Farmers, herdsmen, stock-breeders, and the like.
7. Chiefly Manual	312	5,420,899	Laborers on farms, 3,323,876; laborers not specified, probably in part on farms, 1,857,023; miners, 240,000.
Total	1,000	17,392,899	

proportion whose purchasing power is not on the average above that of a first-class mechanic. In Class II one half of the number consists of clerks, salesmen, saleswomen, and other minor employés, whose purchasing power would stand between that of a factory operative and a good mechanic.

Subdividing these two classes, we then have among every 1,000 in purchasing power—

GRADE I.

Persons of high purchasing power—

Class I, one half	20	
Class II, one half	30	
To these may be added perhaps one fifth of Class VI, prosperous farmers	50—	100

GRADE II.

Medium purchasing power—

Class I, teachers, etc., one half	20	
Class II, clerks, etc., one half	30	
Class III, factory and machine-shop operatives, all	100	
Class IV, mechanics, all	107	
Class VI, two fifths of the farmers	100—	357

GRADE III.

Lowest purchasing power—

Class V, servants, etc.	131	
Class VI, two fifths of the farmers	100	
Class VII, laborers	312—	543
Total		1,000

Factory operatives are classed in Grade II, because their food is usually purchased with intelligence at low prices.

From this analysis it will appear how much the activity of trade may depend upon the purchasing and consuming power of Grade III, numbering more than one half of the whole working force.

The greater part of what is called "the business of the country" consists in the exchange of the necessities of life. The difference in the actual consumption of food, fuel, and clothing between the rich, the well-to-do, the mechanic, the operative, and the laborer, consists more in quality and method of service than in quantity, and therefore any lack of occupation which deprives a large number even of common laborers of their customary supply of such articles will affect the trade of the merchant, the traffic of the railway, and the sale of the products of the manufacturer in vastly greater measure than a temporary commercial crisis which only changes the ownership of realized wealth. The present period of depression must be considered in this light ; it is very different from the ordinary commercial crises such as those of 1836 and 1857.

Let it be remembered that in 1882 about 650,000 men, mostly laborers, were employed in the mere construction of railroads, and that in 1884 not exceeding 220,000 were occupied in this work. Let it next be remembered that in 1880, 1881, and 1882 there was a rapid and progressive increase in the number of factories and works of all kinds, which had almost wholly ceased in 1884, from which cessation perhaps not less than 250,000 men must have been thrown out of employment. In these two facts we have evidence of lack of customary employment for about 680,000 men, or nearly 8 per cent. of the whole consuming force in Grades II and III.

Whenever this partly idle force shall have been placed on new land or found new work, or whenever confidence and capital begin to work together on the old lines, the present depression may end and prosperity may be renewed.

It is the common laborer who suffers most in a period of depression, and if I am even approximately correct in my estimate of the number of laborers discharged by the cessation in the construction of railways, mills, and works, not less than one quarter of all the laborers not listed as on farms have thus suffered. In a true diagnosis we must find the seat of the disease before we can apply the remedy.

Only approximate accuracy can be claimed for the analysis contained in the foregoing table, because the groups are not capable of absolute definition. It is suggestive rather than conclusive. For instance, all that were occupied as machinists, on clothing, boots or shoes, milliners and the like, have been placed in the collective factory work, because such is the tendency of these arts ; but many such persons belong in the mechanical group, not collective but individual.

On the other hand, many of those placed in the latter class are doubtless connected with large factories. The doubt having been given to the collective factory group, which actually counts only 92 to 93 instead of 100. This classification may be accepted as fairly accurate, and it shows a somewhat surprising result. It proves how little we have yet displaced handwork and individual faculty or *gumption* by the substitution of automatic machinery. The improvement in the tools which are guided and directed by hand and brain has perhaps been much greater than the substitution of automatic machinery.

There is less uncertainty in regard to the other groups. One thing is very certain, and that is by far the greater portion of all who are occupied in any gainful work are in the position of wage-laborers or small farmers, and therefore any cause of depression which impairs their purchasing power by lack of employment, or by reducing their wages or earnings, must react with very great severity upon the profits of manufacturers and merchants. So far from the interests of the several classes being antagonistic, they are interdependent, and there is nothing so adverse to high profits of capital as low wages for labor. Where is the remedy ?

The period of depression through which we are passing is very similar to that which ensued after the so-called panic of 1873, and may find the same remedy, unless the world is really overstocked with the products of agriculture ; a condition which at any rate cannot last long. Given a demand for grain, meat, and dairy products, the land still offers relief, and it is in a redistribution of laborers upon new land that relief must soon come.

It will be observed that in this, as in other periods of depression, the sales of government and of railroad lands have been very large, but the change in the distribution of labor by a transfer to new land is very slow, and a long time elapses before new settlers become large consumers of manufactured goods. People continue to wear old clothes when out of work or when changing their mode of life

The cessation of constructive enterprise in 1872-73 was very sudden, the redistribution of laborers afterward was very slow, but by January 1, 1879, when the specie standard was restored and all doubt ceased for a time as to the stability of the currency, every condition was ripe for the activity and prosperity which ensued.

In the same way, and in even greater measure, after the excessive railway construction of this decade culminated in 1882, the cessation of constructive activity in all directions was sharp and severe, but since then the redistribution of labor has been steadily progressing, and if all doubt as to the stability of the currency could be again removed by the cessation of the coinage of silver, a period of activity and prosperity might quickly come.

Our population is now gaining with great rapidity, and the absolute demand for shelter, clothing, subsistence, and additional means of communication for this increase cannot be long held in abeyance. The country is full of all the elements of wealth, and just as the restoration of the specie standard in 1879 gave the necessary confidence then, so might the cessation of the coinage of silver dollars give confidence in the stability of the standard of value now, so that the activity and prosperity of 1880 might recur in 1885 if Congress would act at once.

The number of persons out of work at any given time is always exaggerated, because common laborers, when out of work, always flock to the city in search of employment, and, being thus concentrated, appear to be in greater force than they really are ; yet, when even a small percentage of labor is idle, it has the same effect on the general market for labor that a small excess of goods has on the market for goods. One adverse condition reacts upon the other rendering both more intense until the time arrives when constructive enterprise can no longer be deferred, then consumption is renewed at its normal rate.

The only question now is: Has the time arrived in 1885 for preparation to be made for the increase of population of about 2,000,000 in 1886?

Even in 1884 we have found it necessary to add 4,000 miles of railway. Shall we need 6,000 presently in one year? If so, over 100,000 idle men will be set to work on 2,000 miles of additional track.

Will each family of five in the increase of 2,000,000 require a house, or part of a house, and furniture at an average cost of $500 per family? If so, the work of 500,000 mechanics and laborers will be needed at $400 each per year to supply them.

What other provision must be made? Each one can reply according to his judgment. Suffice it that at a certain date, sooner or later, constructive enterprise must begin, and when it does every man now idle will be set to work and many more will be needed. With an excess of capital waiting to be invested and an excess of labor waiting to be used, and with a peremptory necessity for constructive work near at hand, what other cause can be assigned for continued commercial depression, *except the uncertainty as to the standard of value which is caused by the coinage of low-priced silver dollars?*

The utter insignificance of the silver product as compared to others is shown by the accompanying table. I have called the silver interest a "fly-speck." Are the conditions now ripe for prosperity to be retarded in their beneficent action in deference to a political and economical "fly-speck"?

It is possible, by a graphic comparison of the annual value of our product of silver with those of food, clothing, and other staples—at present reckoned on the gold basis of 100 cents to the dollar,—to give a clearer notion of the confusion into which the business of the country is sure to be thrown if the act for the enforced coinage of silver is not soon repealed.

The following table gives the relative value of the silver product of the United States, shown by a comparison with some other important articles of consumption. The value of the articles of food given is on the basis of the average consumption of each person in the United States (counting two children under ten as one person) being assumed to be equal to the ascertained consumption of cotton-factory operatives in New England and in the Middle States. The estimates of the value of clothing and other articles made from fibres, and of cotton, wool, and iron, are approximate, but sufficiently accurate for purposes of comparison. Population reckoned at the consuming power of 50,000,000 on a probable total population of 57,000,000, counting two children under ten as one adult.

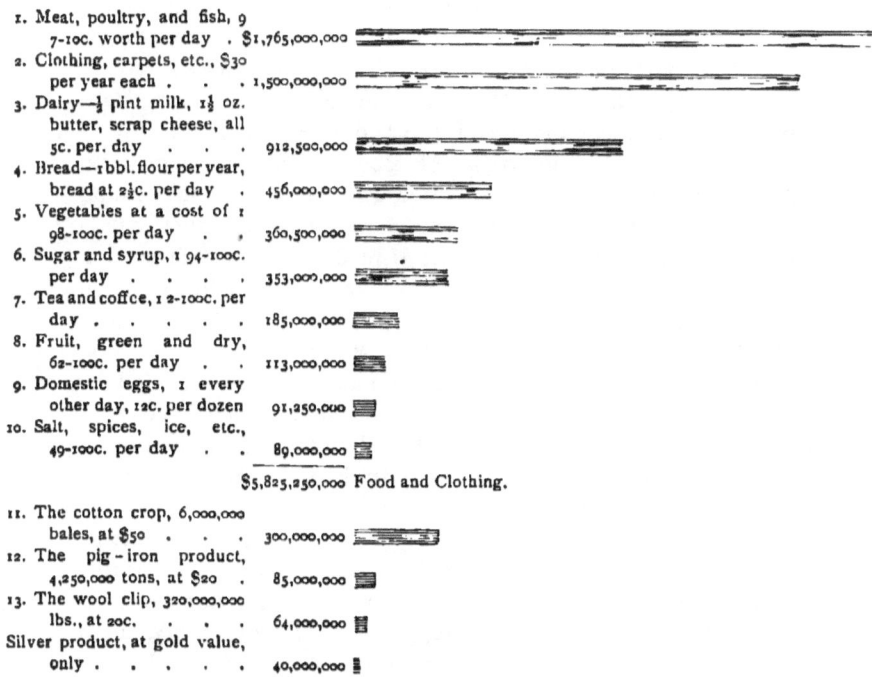

1. Meat, poultry, and fish, 9 7-10c. worth per day . $1,765,000,000
2. Clothing, carpets, etc., $30 per year each 1,500,000,000
3. Dairy—½ pint milk, 1½ oz. butter, scrap cheese, all 5c. per. day . . . 912,500,000
4. Bread—1 bbl. flour per year, bread at 2½c. per day . 456,000,000
5. Vegetables at a cost of 1 98-100c. per day . . 360,500,000
6. Sugar and syrup, 1 94-100c. per day . . . 353,000,000
7. Tea and coffee, 1 2-100c. per day 185,000,000
8. Fruit, green and dry, 62-100c. per day . . 113,000,000
9. Domestic eggs, 1 every other day, 12c. per dozen 91,250,000
10. Salt, spices, ice, etc., 49-100c. per day . . 89,000,000

$5,825,250,000 Food and Clothing.

11. The cotton crop, 6,000,000 bales, at $50 . . . 300,000,000
12. The pig-iron product, 4,250,000 tons, at $20 . 85,000,000
13. The wool clip, 320,000,000 lbs., at 20c. . . . 64,000,000
Silver product, at gold value, only 40,000,000

The above ration of sugar, tea, and coffee of the factory operatives is, doubtless, considerably above the average of the whole country ; but the ration of food taken, as a whole, is not a very large one, as will be seen by a reference to the items in the preceding treatise. This table is based on the statistics of the food consumed by adult women chiefly ; men consume a larger ration of meat and less tea, coffee, and sugar.

It will be observed that the three products which claim special legislation most urgently, to wit : pig-iron, wool, and silver combined are worth only $189,000,000, which is less than the value of poultry and eggs, and but a small fraction of the value of the products of the dairy in each year.

The value of the total consumption of the United States may now be computed (1885) at about $11,400,000,000. The graphical line representing it would be nearly six and a half times the upper line shown in the table, with which the fly-speck which represents silver may be compared. More than one half of the silver production is purchased by the Treasury for coinage.

The foregoing list of articles of food and clothing amounts to $5,800,000,000 (omitting raw cotton and raw wool, and treating pig-iron separately). It represents a somewhat less sum than is probably paid by the people of the United States for such articles. The basis of the table, so far as food is concerned, is on the standard of the actual consumption of factory operatives, chiefly women, at a cost of 23 $\frac{85}{100}$c. per day, or $1.67 per week. It is probable that the average cost in money of the food of adults is more than this, although it is not probable that they average as good a ration for their money, the food of these operatives being bought at wholesale prices. Food, drink, and clothing cost the consumers of this country about $6,500,000,000 per year on the basis of the present population. Pig-iron, when converted into its final form of bars, rails, castings, bolts, nuts, and the like, probably adds $300,000,000, and there still remain timber, stone, and all material for shelter to be added. As I have stated the value of all the products of this country at this time is probably over

$11,400,000,000—or, deducting the domestic consumption of farmers, our commercial product at the point of final consumption is worth over $10,000,000,000; but, it must be remembered that in the process of exchange and of conversion this whole product will have been bought and sold twice, thrice, or more times. Before it reaches the consumer the wheat has been sold by the farmer to the miller, the flour has been sold by the miller to the merchant, and by the merchant to the baker, and the bread has been sold to the consumer. The business transactions—the purchases and sales of this country—must approximate $30,000,000,000, or between five and six hundred dollars a year per capita, in the mere transactions relating to shelter and subsistence.

Whatever the final amount may be, the prices are now adjusted to the standard of the gold dollar, rated at 100 cents.

When the standard is changed to silver at 82 cents to 85 cents, as it surely will be unless the coinage of legal-tender silver dollars is soon stopped, the prices of this immense volume of consumable commodities as well as of all other property not enumerated, must rise in just the proportion that the standard of value is lowered. This rise will be very slow, because consumption has been so much reduced by uncertainty. The probabilities are that while this adjustment is in process wages will keep where they are or go lower, while the money cost of living will become greater. In such periods the rich grow richer at the cost of the poor, but the principal loss falls on the persons of moderate means.

The absolute necessity of preparation for an increasing population may counteract these tendencies in a measure, but no enterprise or vigorous activity will be possible, and, on the whole, depression and want of work will be continued, in the face of rising prices and increased cost of subsistence.

The legislators who sustain the present acts of coinage, which are approved neither by bimetallists nor monometallists, will be responsible for the disturbances which will ensue.

Having thus considered the distribution of occupations with reference to the Railway Service and the Silver Coinage, we now come to the apparently more complex but really much more simple question of

THE COLLECTION OF THE NATIONAL REVENUE.

It is held by one school that domestic industry will be protected by the imposition of duties under a tariff in such a way as to raise the price of such foreign articles as can be made in this country, in order that various arts may become established which it assumed might otherwise be of very slow growth, or perhaps might not be undertaken at all.

It is held by another school that domestic industry will be most fully promoted by levying duties or taxes exclusively on articles which are of voluntary use, which do not enter directly into the processes of domestic industry, or which cannot be produced in this country advantageously, if at all.

It is not the purpose of this treatise to discuss the merits or demerits of either system but rather to define the necessary conditions, and to state the facts which must be accepted by the respective advocates of both systems, if any scientific result is to be reached.

Three questions are presented to which a sufficient answer can be given by an analysis of the table of occupations.

1st.—What proportion of the gainful occupations of the people must be carried on within the limits of our own territory because they could not be conducted as well elsewhere?

2d.—What proportion of all who are occupied depend upon a foreign market for the sale of their product?

3d. What proportion of all who are occupied could be subjected to foreign competition?

Agriculture is the most important of all occupations. In the census year it gave employment to 7,670,493 persons, and probably to a greater number, as the Superintendent of the Census remarks that many of those who reported themselves simply as laborers were probably *farm* laborers. The census does not indicate the proportion of persons to each special crop, but if consideration be given to the estimate of the crops at their farm values, the total product of agriculture possessed a value in the census year of $3,726,331,422 ; or, with transportation added to the place of export, or of *wholesale* distribution, the total approximated $4,000,-000,000 in value. The declared *wholesale* value of the products of agriculture exported in the same year was $685,961,091, which is 17 $\tfrac{16}{100}$ per cent. of the whole. At the present time it would be somewhat less. If we apply this percentage to the whole number of persons listed specifically as occupied in agriculture in the census year it gives us 1,315,000 persons engaged in domestic agriculture whose market was a foreign one. If the number occupied in agriculture was greater, then this number must be increased.

On the other hand, we find among the products of agriculture which could be wholly or in any substantial part imported—only sugar, swamp rice, a part of the wool, tobacco, barley, and hemp, and a few minor articles,—the possible import of what was produced here not exceeding $100,000,000, or two and a half per cent. Applying this percentage to persons, we get 192,000 whose occupations might be affected by changes in the revenue system.

If we next consider the several classes of occupation, aside from agriculture, we find that the all of Class I, who were engaged in professional work, or as officers of railroad, insurance, and other similar corporations ; all of Class II, who were engaged in distributive work, as merchants, traders, and their employés ; substantially all of Class IV, engaged in mechanical work of the individual rather than of the collective kind ; and all of Class VII, laborers and miners, with the exception of about 32,000 iron miners and 20,000 coal miners supplying blast furnaces, must have lived and worked within the limits of the country, and in such parts of the country as are consistent with the vocation of each individual, because their work could not be done elsewhere.

There remains Class IV, comprising at the utmost 1,740,000 persons, engaged in collective factory work. Of this number a large portion of those who were engaged in metal and machine work, almost the whole number employed in making clothing, boots, shoes, and hats, and by far the largest portion of those engaged in the lesser branches of collective factory work, such as wood-working, and other kindred arts, must have followed their work not only within the limits of the country itself, but in such particular part of the country as was best suited to their special work. Of the whole number of this class, computed at 1,740,000, possibly 740,000 might be in part subjected to competition from a foreign country; to whom may be added 260,000 agriculturists and miners, making 1,000,000 in all.

Each person's judgment would vary somewhat as to the proportion of the persons engaged in manufacturing, mining, and agriculture, whose product could be imported at this time if no discrimination were used in the imposition of duties, but it is impossible to reduce the problem to absolute terms. So long as duties are imposed on ores, coal, wool, chemicals, and other articles, which enter into the processes of domestic manufactures, the import of articles made of iron, cloth, and other finished articles will be greater. The proportion of all persons occupied who can be subjected to foreign competition may be estimated at between four and six per cent. of the whole; a proportion which represented between 700,000 and 1,050,000 persons in the census year.

Paying no regard to the small proportion of domestic manufactures exported, the general result appears to be that in the census year 1,300,000 to 1,350,000 persons occupied in agriculture depended upon a foreign market, and from 700,000 to 1,050,000 were occupied in some kind of production which could have been imported wholly or in part. Assuming the maximum in each case, we find, in round figures, 2,400,000 persons employed whose occupations were directly connected with or affected by foreign commerce. The remainder of the working force, 15,000,000 in number, living within our limits, were, of necessity, occupied in kinds of work which could only be done within the same limits. Hence the vast and necessary preponderance of domestic over foreign commerce.

It will be apparent that these conditions which exist in the nature of things must be fully comprehended before any intelligent legislation can be had in respect to national taxation, whether the revenue is to be sought either under an excise or from a tariff.

The importance of our foreign commerce is not, however, to be measured by the ratio which it bears to domestic traffic. Possessing as we do the most adequate resources, and the cheapest, because the most effective, labor of the world, we are enabled to supply our own wants, and yet produce an excess of staples which the world must have. Hence it follows that imports and exports constitute the balance-wheel by which the price of our whole product might be maintained more uniformly than it is, were it not for the obstruction of ill-adjusted taxation. The effect of these obstructive duties upon the import of articles which enter into the processes of domestic industry is to increase the general cost of our product, and to reduce its exchangeable value ; hence it follows that the general rate of wages is lower than it would otherwise be, and is also subject to unnecessary fluctuations.

Under the present complex and onerous tariff, which discriminates in many ways against our domestic manufactures, a larger proportion of those who are occupied in them are subject to foreign competition than would be the case under a well-adjusted tariff.

If all the materials which enter into the processes of domestic industry, commonly called raw materials, were free of duty, as well as finished products which are necessary thereto, such as chemicals, drugs, and dyestuffs, the number of persons who could be subjected to foreign competition, by way of importations of manufactured products of like kind, would not exceed about 500,000, to whom may be added not over 200,000 in agriculture, mining, and metallurgy. Such a policy would, on the other hand, greatly promote the export of manufactures as well as of the products of agriculture, and in this way would increase the general rate of wages by widening the market, and thereby enabling the country to obtain a larger sum of money for its excess of production. The interest of every machine-using nation, in which wages are naturally high, is to get the benefit of the cheapness of its highly paid labor by opening the widest market by the exchange of its goods for products made under less advantageous conditions, and, therefore, at low rates of wages.

This benefit can only be fully attained when industry is untaxed.

Duties upon finished goods which are ready for final consumption rest upon an entirely different basis. They may be so imposed as to yield a large revenue without any material obstruction to industry beyond the amount of the revenue itself, and it is in this adjustment of duties and taxes that the most careful discrimination is required ; but this branch of the subject is foreign to the purpose of this treatise.

The conditions of industry in the United States are very different from those of almost any other country, because there is no article necessary to subsistence which we cannot produce in ample measure, if we choose to do so. In making this statement, tea and coffee are placed among the comforts rather than the necessities of life ; aside from these we could produce every thing of any considerable importance. It may be great folly to undertake to do so, because the conditions under which sugar, iron, jute, and many other crude articles are produced are very arduous and undesirable, and in some cases unwholesome. When such articles can be procured by exchange at a lower cost than by their domestic production, the advantage lies with the country which is not compelled to do such work. The same rule holds true with respect to many articles of a high grade in which the labor is mostly ill-paid hand labor. We cannot afford to spend our time on such work when the very poor and ignorant of other countries can do it so well for us, and can do nothing else for themselves.

Hence it follows that the measure of our imports and exports is rather the out-come of our abundance, while in Great Britain it is the measure of her necessity, since her people could not be subsisted except for her commerce with other lands.

The obstructions which we have interposed to the import of materials which enter into the processes of domestic industry, such as coal, iron, salt, hemp, jute, chemicals, dyestuffs, timber, etc., give a great protection to the manufacturers of Great Britain so long as these duties keep the prices higher in this country than there. It matters not what the absolute price may be, whether high or low, so long as there is an artificial difference against us, we lose the benefit of our more effective labor and give this benefit to Great Britain, her labor being more effective and her wages higher than any other competitor on the continent of Europe. It is not to be wondered at that Great Britain views with alarm any change of policy in the United States which will bring us into direct competition with her in her foreign markets.

The productive power of this country can be more adequately proved by an analysis of the work of a single State.

The State of Ohio has been taken as an example more than once in the course of these treatises. It lies midway between the East and the West, and far enough North to be in the temperate zone, most conducive to success in manufacturing enterprises. It possesses great resources, both in respect to agriculture, mining, and manufacturing. Disregarding fractions, the proportions of its populations who were engaged in gainful occupations were as follows:

Total number in all occupations, 994,475, or one in each 3.21 persons, against an average one to 2.90 in the whole country.

Of this number of persons there were engaged in agriculture, 397,495; in professional and personal service, 250,371; in manufacturing, mechanical, and mining occupations, 242,294; in trade and transportation, 104,315.

Again, disregarding fractions, the proportions were almost the exact average of the whole country, to wit: 40 % in agriculture; 25 % in professional and personal service; 24¼ % in manufacturing, mining, and mechanical work; 10½ % in trade and transportation.

If we analyze the work of these several classes, in order to determine in what measure the people of Ohio could be subjected to foreign competition, even including the competition of the adjoining Dominion of Canada, the result may be somewhat surprising.

First, with respect to agriculture. There is probably a little import of barley into Ohio from Canada for the purpose of making beer. There may be some interchange of agricultural products, of fruit and the like; and perhaps a little exchange of Ohio spring wheat for Canada winter wheat. But there is no crop of any substantial importance raised in Ohio which could be subjected to a serious foreign competition, except wool.

The total value of all the products of agriculture in the State of Ohio in the year 1883 was computed by the State Commissioner of Agriculture at somewhat over $184,000,000—which would be substantially at the average rate of product to each person occupied in agriculture which has been assumed throughout this treatise, *i. e.*, a little over $400 per year.

The wool clip of the present year is computed at 24,000,000 lbs., worth about $7,000,000, or about 4 % of the whole product of agriculture. At this ratio, assuming that each person raising wool did nothing else, the proportion of those who are engaged in agriculture who depend upon wool for their subsistence would be not over 16,000 in number.

In point of fact a few sheep are kept by many farmers, and very few persons, except the breeders of high-priced rams for breeding purposes, depend in any large measure upon sheep-growing or the wool clip.

In its place Ohio wool is about the best of its kind, and it could not probably be displaced by any possible importation from any other country; but, assuming that it were thus displaced, it would affect the employment of the people in the proportion of one person in twenty-five of all who were occupied in agriculture, or of one person in sixty-two of all who were occupied in all employments, assuming that it were their sole occupation.

The persons engaged in professional and personal service and in trade and transportation in Ohio cannot, of course, be subjected to foreign competition.

We may therefore consider manufacturing, mining, and the mechanic arts by themselves.

In the census year, the entire number of persons engaged in the production of iron and steel within the limits of the State of Ohio, was a little under 20,000, or two per cent. of all who were occupied in gainful occupations. A few other branches of industry might be subjected to foreign competition, but the whole number in all branches of mining, mechanical work, and manufacturing could not exceed 25,000. How many of these could be absolutely displaced only time and experience could prove. But assuming that the occupation of this whole number were of necessity altered by foreign competition, it could only happen for the reason that the people of Ohio could procure more iron, steel, and glass from some other country by an exchange of products therefor ; it would follow of necessity that by so much as these arts were given up, some other arts would be undertaken, because the people must have iron, steel, glass, and other like commodities, whether produced by themselves or by foreigners. If they did not produce these articles themselves they must produce something to exchange for them.

Summing up all products of agriculture and all products of mining or manufacturing which can be imported into Ohio from a foreign country, we find that foreign competition would be limited to less than 4 to 5 % of all engaged in gainful occupations, while the other 95 to 96 % live and work within the State of Ohio because there is no other place in which, in their judgment or in fact, their work can be done so advantageously to themselves.

Reference may be made to the details of the occupations of the people of Ohio in the preceding treatise.

But more significant are the details of the occupations of the State of Pennsylvania, whose people resist the remission of taxes on the materials which are most necessary in all arts, industries, and occupations, whether of agriculture, the mechanic arts, or manufacturing, more urgently than the people of any other State.

OCCUPATIONS IN PENNSYLVANIA, 1880.

Agriculture		301,112
Professional and Personal Service		446,713
Trade and Transportation		179,965
Apprentices	8,907	
Bakers	6,025	
Blacksmiths	20,276	
Bookbinders	2,055	
Boot and Shoe Makers	20,634	
Brewers	1,504	
Brick and Stone Masons	16,210	
Brick and Tile Makers	4,504	
Butchers	9,200	
Cabinet-Makers and Upholsterers	6,866	
Carpenters and Joiners	40,782	
Carriage, Car, and Wagon Makers	6,026	
Cigar Makers and Tobacco Workers	8,970	
Coopers	3,852	
Engineers and Firemen	11,452	
Fish and Oysters	598	
Gold and Silver and Jewellers	2,204	
Harness, Saddle, and Trunk Makers	3,729	
Leather Curriers, etc.	6,020	
Lumbermen, etc.	4,085	
Machinists	14,601	
Millers	5,902	
Painters and Varnishers	13,008	
Plumbers and Gas-Fitters	2,621	
Printers	7,877	
Saw-Mill Operatives	4,619	
Tailors, Dressmakers, and Milliners	49,851	
Tinners	5,264	
Wheelwrights	2,381	
Miscellaneous arts, each small in number	67,561	357,584
Substantially exempt from foreign competition		1,285,374
Subject in part to the competition of a product of like kind, which could be imported from a foreign country:		
Clerks and Book-Keepers, Manf'g Co's	1,668	
Cotton, Woollen, and Silk Mill Operatives	44,746	

Employés, Manf'g Co's not specified	3,995
Iron and Steel Workers	33,628
Manufacturers and Officials, Manf'g Co's	6,740
Mill and Factory Operatives, not specified	6,701
Miners	69,415
Paper Mill Operatives	2,176
Ship Carpenters, etc.	1,624
	170,693
Total	1,456,067

In the foregoing classification it may be admitted that a small portion of those who are listed as blacksmiths, carpenters, and who are engaged in transportation, and in the miscellaneous list covering glass, chemicals, etc., should be added to the list of persons subject in part to foreign competition ; but on the other side all clerks, etc., of manufacturing companies, all manufacturers, mill and factory operatives not specified, and officials of manufacturing companies have been placed in the list of those who are subject in part to foreign competition. Foreign competition is therefore narrowed down to about 150,000 persons engaged in mining, metallurgy, textiles, paper, glass, and chemicals.

The proportion of miners engaged on iron ore and coal for blast furnaces is less than one half the whole number. A large portion of the cotton and woollen fabrics are coarse goods which can be made here at a lower cost than in Europe, and a large portion of the iron and steel workers would have more work rather than less, if iron ores, pig-iron, and ingot steel were free of duty or tax.

It would be difficult to prove that more than one half the list of those whose industry might be subject in part to foreign competition would be for a time adversely affected, even if a policy exempting the materials which enter into the processes of domestic industry from duties was adopted, such as pig-iron, coal, wool, timber, chemicals, salt, and the like, while all other arts would be promoted. One half would number 85,000 or less than six per cent. of the whole number of persons occupied in all gainful occupations in this State.

It therefore follows that so long as a tax is continued upon the import of a foreign article which is needed in the processes of domestic industry, by which the price of that article, however low it may be, is yet kept higher than it is in Europe, the manufacturers of this country are kept at a relative disadvantage, and perhaps no art suffers so much from this cause as the art of shipbuilding on the Delaware River in Pennsylvania itself.

It indicates a singular delusion when out of 1,456,067 persons occupied in all gainful employments in the State of Pennsylvania in the census year, not over 30,000 to 35,000 were employed in mining iron ore, in mining coal for blast furnaces, and in the conversion of these materials into pig-iron. It should yet appear from the public utterances of the public men of this State, as if the people were incapable of sustaining themselves if this undesirable occupation were not specially promoted.

Pennsylvania possesses agricultural resources unequalled in this country, timber, oil, fuel, power, great navigable rivers, and every other advantage which nature can give her ; but yet subjects them all to a grave disadvantage in order to attempt to sustain, by purely artificial and obstructive methods, a branch of work which is not desirable in itself in its necessary conditions, and which is now being subjected to a destructive domestic competition, perhaps prematurely forced into action by the very policy which she herself has insisted upon.

Although it is not the purpose of this treatise to enter into the general discussion of the question of taxation, yet it has become apparent that no treatise upon the forces which make the rate of wages can be considered complete which does not take cognizance of the taxation imposed upon coal, wool, timber, and pig-iron, whereby this country is placed at a relative disadvantage compared to almost all others.

Such taxes upon the very sources and foundations of industry cannot fail to reduce the rate of wages by restricting the sale of our products of other kinds while at the same time increasing their cost.

No determination of other questions is possible until the power of the representatives of pig-iron, wool, and silver to dictate the policy of both political parties is taken away, nor until their influence is reduced to the measure of their importance. Their relative importance is easily measured by a very commonplace standard.

Their gross annual value is now at the maximum less than $200,000,000, or less than two per cent. of our annual product.

Each one taken by itself represents a less product than the product of eggs from the hen yards of the country, and the three together barely equal in value the product of eggs and poultry combined.

So long as their domination is submitted to, the adjustment of the tariff is impossible. No advocate of free trade can ask a heavy reduction of duties on fabrics which are ready for final consumption, when the materials of which they are composed are subject to excessive duties ; and no advocate of the protective policy can make even a reasonable concession so long as manufacturers of iron, steel, woollens, worsteds, and other fabrics are subjected to such a burden as the present tax on materials. No determination can be reached as to what is the true or possible maximum rate of wages in this country so long as all our workmen are placed at such a disadvantage as is imposed upon them by heavy taxes on the most necessary articles which enter into the processes of their industry.

Entirely aside from these temporary questions of currency and taxation, we may again question the table of occupations to see why the average production is so small as 50 to 55 cents per day per capita, or $1.45 to $1.60 per day to each person occupied in gainful work ; and also why, small as it is, it is so unequally distributed.

What has this inequality to do with the alleged monopoly of land by private owners which is said to exist by Henry George and other sincere reformers of the same school.

For this purpose we may limit our consideration to the United

States, where the purchase and sale of land has been made more simple and free from legal obstruction than in any other country, except some of the colonies of Australia where there is reason to suppose that even a better and more simple mode of sale and transfer of land exists than with us. The fault, if any, in the system of private ownership cannot be determined by a study of the condition of Ireland,—a small island which is still subject to the disabilities caused by despoiling private owners under the alleged right of conquest long years since ; nor by a study of English land, burthened as it is by rights of dower, settlements, and entails to such an extent that actual ownership of the larger part of the soil has practically ceased to exist, most of it being held under a life estate only ; nor by a study of the conditions of most of the continental states of Europe, where compulsory subdivision of land has in great measure prevented the wide application of capital to its most productive use.

In this country a very large portion of the soil has been and still is under State or National ownership ; it does not need to be *Nationalized*, because it is *Nationalized* already. It has long been practically free and open to homesteaders, preëmptors, squatters, graziers, ranchers, and the like, and all our efforts have been to get it into private ownership or occupancy, in order that it might be put to productive use.

Even a large portion of the land held in private ownership has been and is practically open to occupation and use at so small a price as to be substantially *free* land. A large portion of the mountain section of the South, unequalled in its potentiality for production, or in natural conditions favorable to health and industry, has been and may still be purchasable at from twenty cents to two dollars per acre in fee simple.

It follows that if there is want in the midst of abundance, and if the poor of our cities are crowded into slums, it is not to be attributed to lack of free land. In fact, we waste the powers of the land that is in use, for the mere reason that there is so much of it not yet occupied for use, and this wasteful method may be defended as the most economical for the time being. Again the graphical method may be employed to make this matter plain.

In the following table I have dealt in a rough and ready way with the areas occupied, or which might serve, for all our great crops. The area of the United States, omitting Alaska, is a trifle less than 3,000,000 square miles.

In a broad and general way we may assume that one half this area is good arable land, one quarter good pasture land, and one quarter forest, mountain, and mining territory.

TOTAL AREA.

3,000,000 square miles

Graphically shown by the four lines.

Mountain and Timber.
1-4

Grazing.
1-4

Arable.
1-2

INDIAN CORN FIELD.

112,500 square miles.

At 25 bushels to an acre this area produces 1,800,000,000 bushels. This corn is largely converted into pork at the rate of 5 lbs. of corn to one pound of pork. Assuming one thousand million bushels thus converted, and the rest used for human or cattle food, the product of pork would be equal to 18,500,000 casks or its equivalent in bacon ; which would give nearly one cask of pork of 300 lbs. to each head of a group of three persons per year, or 100 lbs. per capita.

WHEAT FIELD.

60,000 square miles.

At 13 bushels per acre this little area yields a little over 500,000,000 bushels. Setting aside an ample portion for seed this quantity would give over 80,000,000 persons one barrel of flour per year.

COTTON FIELD.

20,000 square miles.

At the wretched average of only half a bale to an acre this little patch yields 6,400,000 bales in a year.

WOOL.

What the actual area of sheep pasturage is no man can tell, because the area of land absolutely free to graziers and ranchers is so large that no question of area has arisen until within a very short time; but the end of this wasteful and archaic method can be foreseen. When the cur-dog shall have been muzzled, or when dogs shall have been declared *feræ naturæ*, it will be easily possible to sustain four sheep to an acre over wide areas of unoccupied land in the East and South as well as in the far West; this would require a sheepfold of

40,000 sq. miles,

sustaining 102,400,000 sheep, which at only 4 lbs. each would yield more wool than we now consume of all kinds both domestic and foreign.

DAIRY FARMS AND HEN YARDS.

In 1880 the number of milch cows was estimated at 12,500,000, and the product of eggs was computed at 500,000,000 dozen, valued at $80,000,000. Over how wide a range of pasturage the milch cows ranged it is impossible to say, but almost within the period which has elapsed since 1880 it has been proved entirely possible to feed two cows one year on the corn-stalks saved in pits which can be raised on one acre of fairly good land, if to this green fodder be added a ration of meal made from the cotton seed which was almost all wasted until a very recent time and is yet saved in only a very small proportion. But in order to be safe we may reverse this ratio, and assigning only one cow to two acres we may greatly increase our present ration of milk, butter, and cheese, with the hens' eggs thrown in.

A Dairy Farm and Hen Yard
of 60,000 square miles,
at 1 cow to 2 acres, will sustain 19,200,000 cows.

BEEF.

The relative importance of meat in the subsistence of our people has been shown in the foregoing table. A large portion of our beef is now produced by almost semi-barbarous methods on the far-distant plains ; but as population increases this rude way must give place to more civilized and humane modes, and our beef must be produced near its place of consumption. Many Eastern farms which had ceased to be profitable have lately been converted into *beef factories*, upon which steers are raised and fattened on ensilage and corn-meal. Provision has been made for the cornfield, and if pitted forage is as fully justified on a broad scale as it has been in the successful experiments of many able men who have applied brains and capital to the use of land, it would be necessary to assign only a small area to beef.

60,000 square miles,

at 500 lbs. of meat to an acre, would yield nearly one pound of beef per day to our present population (reckoning two children as one adult).

If these propositions can be sustained, it follows that our present crops of corn, wheat, and cotton, and a very much increased product of the dairy and poultry-yard, as well as of meat and wool, can be raised on

352,500 square miles,

or upon twelve per cent. of the total area ; and even this assignment of land is nearly double what might be required if the *intensive* system of farming were adopted by men of sufficient intelligence and capital to conduct all parts of the work in a reasonably good way.

It is held that in the face of this demonstration the charge that poverty is now to be attributed to monopoly of land in this country is utterly disproved, and that the explanation of extreme poverty must be sought in other directions. It is painfully apparent that extreme poverty is to be found chiefly among those who are foreign born, but there is as much free land open to them as there is to the native born—enough and to spare for both.

It may therefore well be questioned whether the more intense and widespread poverty of European countries can be attributed mainly, even if in part, to the systems of land tenure there prevailing, if the same phenomena are to be found in the heart of the great cities of this country where there is so much free land as in those of countries where land is fully occupied.

Want oppresses New York, Boston, and Philadelphia, of the same kind if not in as great a degree as London, Paris, and Berlin. Yet in the United States land is in excess of the utmost need; in England it is held by the few rather than the many, and in France and Germany by the many rather than by the few. It follows, almost as a matter of necessary deduction from these phenomena, that the great problem is the distribution of the product of the soil rather than the distribution of the soil itself. The greater part of those who only suffer in cities might starve if removed therefrom and placed upon unoccupied land where they would depend only upon themselves for subsistence.

What other reason can there be for the very poor to gravitate to the cities, if the struggle to obtain food and shelter is not a little less severe there than it would be in the country?

In order that material welfare may exist at all, labor and capital must both be applied to land. Land is valueless without labor—labor is almost helpless without capital. Is not this the reason why the unemployed flock to the cities where the capital is, and never go to the free land unless moved there and sustained by the capital of others until they can possess land and capital of their own?

I have endeavored to show the function of capital when invested in the mechanism of production and distribution. We may now measure the results of its application and attempt to mark the point which we have reached in our progress toward general welfare.

In the consideration of the following table it must be remembered :

First. That the money-cost of food and drink is probably more than this table gives, because the average working-man buys at retail on less advantageous terms than are obtained by the managers of factory boarding-houses, who buy food at wholesale.

Second. That the estimated consumption of textiles in the form of clothing, carpets, laces, embroideries, and all other forms, is a maximum estimate.

Third. That the estimated cost of shelter for the increase of population is an approximate one only, for which there are but few actual data known.

Proportionate expenditure of the people of the United States in 1884 for food, drink, and clothing, and for additional shelter for the increase of population.

1. Food, at the average ration of factory operatives in New England and the Middle States $4,340,500,000
Drink, as recently computed by David A. Wells . . . 474,823,000

Total $4,815,323,000

2. Clothing ready for use, carpets, blankets, laces, and all other textile fabrics on the basis of the domestic production and import of the census year with the cost of conversion and distribution added $1,500,000,000

3. Shelter for an increase of 2,000,000 on the basis of a dwelling or part of a dwelling for each family of five, costing $500 . . . . $200,000,000

There are no available data for ascertaining the cost of keeping dwellings in repair or of maintaining existing shelter by the substitution of new for old ; but from all the statistics attainable it may be fairly computed that the total value of all the dwellings in existence at this time, or at a given time, for the use and occupancy of all the wage-earners and small farmers of the country, would be little, if any more than the annual market value, at the place of consumption, of the food and drink consumed by them.

In other words, in the foregoing tables I have based all the figures on a population of 57,000,000, equal in consuming power to 50,000,000 adults. In such a population there would be substantially 19,650,000 persons occupied in all gainful occupations, of whom over 18,000,000 would be wage-earners or small farmers, representing at least nine tenths of the actual consumption of the country and sustaining 52,500,000 of the population. If the shelter of each one of these persons is worth $100, the value of working-men's dwellings would be $5,200,000,000, or but little more than the estimated annual cost of food and drink.

Assuming 5 per cent. per annum for repairs and maintenance, we get $260,000,000, which being added to the computed cost of new dwellings, gives the proportion of the cost of working-men's shelter as compared to the cost of food and clothing. How much should be added for rent paid by those who do not own their dwellings would not form a part of this branch of the subject.

What I wish to bring out is this: Out of an estimated product of the present population, at the same ratio as that used in the treatise on wages—to wit, $11,400,000,000 in 1885 against $10,000,000,000 in the census year,

Food and drink take up, at the minimum, about	$5,000,000,000
Clothing, etc., at the maximum	1,500,000,000
Repairs, maintenance, and construction of dwellings for working people	460,000,000
Repairs and construction of dwellings for the well-to-do at double rates	40,000,000
Accounted for	$7,000,000,000

Leaving $4,400,000,000 to be accounted for in the consumption of all other articles aside from food, drink, clothing, and shelter. Out of the distribution of this remainder would come the luxuries of the rich,—the comforts of the well-to-do, and all our additions to capital and to the savings of the people.

But in this analysis it will be observed that the proportion of the total expenditure assigned to food is far short of that which is the well-ascertained proportion in workmen's families both in this country and in Europe, which is fifty per cent. of their income in respect to the better class, and sixty per cent. in the lower grade. At fifty per cent. of our estimated gross income, food and drink costs $5,700,000,000.

What, then, is the conclusion? Is it not that even in this sparsely populated land, of almost unlimited potentiality in its production of grain and meat, more than one half the struggle for life is still a mere struggle for food? Can this low plane of mere existence, which many fail even to attain, be attributed to monopoly of land? to institutions established by law? or to causes wholly remediable by legislation? If not, wherein do we fail, in spite of our much-vaunted civilization?

Again, we must refer to the table of occupations, and in the sorting of all according to their work are we not compelled to admit that a miserably small proportion have become individually capable of making adequate use of the vast resources which have been placed at our disposal? With no lack of land, of capital, of education, or of opportunity, why is it that more than one half of every thousand who are occupied should be found in the position of small farmers working harder than their hired men; or in that of laborers, domestic servants, waiters, and the like? Is not the only remedy to be found in the slow development of individual capacity while the drudgery can only be alleviated by the rapid and safe application of capital?

The wretched hypothesis of Malthus has no place here—neither has it been historically sustained anywhere. Modern sanitary science has curbed the pestilence; famines have become sporadic and of little general effect; war has reduced production in vastly greater measure than it has checked population.

In the light of modern science and experience, a rule might be substituted for this atheistic hypothesis which may be formulated as follows:

Savage man, or even semi-civilized man, while still subjected to the burthen of standing armies and of passive war, tends to increase faster than the means of subsistence can be supplied; under such conditions, pestilence or famine may afford a necessary relief. But civilized man, freed from semi-barbarous conditions, and dwelling in peace, provides means of subsistence in far greater measure than is required by increase in numbers.

Yet, although what may be called the higher laws which make for abundance have only yet been applied in very limited degree, there has not been a decade since the so-called law of population of Malthus was first propounded in which it has not been disproved by a much greater increase in the world's means of subsistence than in the population; subject, of course, to isolated cases where there has been what may be called an artificial congestion of ignorant human beings capable only of being scattered by hunger, as in Ireland in 1846.

Neither has the Ricardian theory of rent nor the opposite theory of Henry C. Carey found any sustaining facts in this country; but a very different formula would be required here. It might be put somewhat in this form:

Rent is the tribute which valueless land renders in proportion to the intelligence, capital, and industry which are applied to its cultivation, use, or occupancy.

If land is devoted to agriculture, the rent which will accrue to the owner who cultivates it himself, or which can be paid by the tenant, will be the produce which is returned by the soil over and above the force expended upon it—which force may consist of labor and capital in varying proportions. The force expended on originally fertile land may be almost wholly labor—upon poorer land it may be almost wholly capital; the measure in terms of money of these two forces may be the same and the value of the product may be the same, while the intrinsic properties or fertility of the soil may have been very different.

If land is occupied for other purposes, rent will be the measure of the advantage of position, or of the efficiency of the capital which is used upon it.

The theory of Ricardo, in the judgment of the writer, is based upon the idea that the soil is a mine, while in fact, it is now treated more as a laboratory ; hence farming has become rather a matter of brains than of muscle. Intelligence and capital rather than labor are now the principal factors in successful agriculture.

It is not, however, my purpose to deal with the views of somewhat insular economists like Malthus and Ricardo, to whom the forces of the railway, the steamship, and of modern chemistry were alike unknown ; nor with those of doctrinaires like Carey, who dwarfed a really observant mind to the petty measure of a purely selfish policy in respect to foreign commerce.

The single question presented to us is this: Have we yet any statistical or historical bases by means of which we can solve the apparently simple problem of

WHAT MAKES THE RATE OF WAGES?

If no fully affirmative reply can yet be given, still great progress is being made. The science of census-taking has been developed in admirable measure by Walker, Wright, and their efficient assistants and coadjutors. Bureaux of the statistics of labor are doing most excellent work in several States and will soon be supplemented by the National Bureau. The work of Mr. Jos. Nimmo, Jr. in the Government Bureau of Statistics leaves little to be desired ; while the State Reports of Railroad Commissioners, supplemented by the Manual of H. V. Poor, give more information on that branch of distribution than can be found in any other country.

In England great progress has been made in statistical science, as well as in Germany and other continental states.

Since the first edition of this book was issued a very valuable report upon " LABOR IN EUROPE " has been issued by the State Department. It is evident that an excellent beginning has been made in the investigation of the condition of laborers in other lands, by American consuls.

The volume just issued gives very full information as to the rates of wages, the cost of food, rents, and other matters which are of the utmost value : but the volume is incomplete, as are most of the reports of the consuls, in not giving any clue to the bearing of these facts upon the cost of production of the most important commodities in the exchange of which this country is interested.

The volume shows conclusively the very much greater share of a larger product which the workmen of this country attain, whether measured in terms of money—*i. e.*, in high rates of wages,—or in what the money will buy ; it also proves that the best conditions, next to this country, are attained in Great Britain, while the scale of wages becomes progressively lower and lower as we pass to the less productive countries of the continent, where longer hours, more arduous conditions, and heavier burthens yield less results in quantity of product and proportion of wages.

Secretary Frelinghuysen's attention has evidently been called to the necessary extension of the work which has been so well begun in his department. In the conclusion of his report he says :

"There are certain natural and artificial conditions which so largely affect the direct conditions of wages as to be entitled to consideration in any analytical examination of the great questions of labor ; but from their abstruseness they are less evident to the general mind and more debatable than the simple relations shown in the reports of the consuls and summarized in this letter. It would be a legitimate field of inquiry to ascertain what are the conditions which enable England to manufacture machinery and other products at less prices than similar goods can be manufactured in France, and at prices equal to those in Germany, *while the rates of wages paid to the workmen engaged in those manufactures in England are, on the whole, higher than those paid for similar labor in France, and more than double those paid in Germany.*"

The italics are my own, and the Secretary might have added : "while the hours of labor are much less per day."

It is greatly to be hoped that this excellent work of the State Department will be continued. It appears that a report by Consul Williams, of Rouen, will soon be given, in which both the rates of wages and the cost of labor in a locomotive engine will be given.

It would not be difficult to frame the instructions to all consuls in such a way that each might report in a similar manner on some given unit.

For instance in Oldham, on the rates of wages paid and the cost of labor on a pound of No. 32 cotton twist.

In Blackburn, on the cost of labor and rates of wages in some described article of woven cotton fabric.

In Yorkshire, Belgium, and Germany, on a 16-oz. cassimere.

In Mid-Lothian, on a ton of wheat.

In Newcastle, on a ton of coal.

In Glasgow, on a ton of iron.

In Germany, on a ton of "basic" steel, or steel-wire rods.

If such reports were accompanied by samples showing the fabric, the mode of preparing for market, and other matters—to be deposited in the Smithsonian Institute,—the reports would leave little to be desired. From these specific statements in regard to certain staple articles, easily compared with our own, the relative cost of all other commodities could be inferred.

For such service high attainments would be required on the part of consuls, which will soon be secured under a reformed civil service, and it is greatly to be hoped that the new administration will give close attention to this most important subject and extend the scope of the work so well begun by the present Secretary.

Since I cannot at present rewrite and thus avoid the repetitions which occur in this volume, it may be well to give the following condensed statement of each of the several conclusions to which I have been led, and which I have endeavored to present and to sustain in the different parts of this treatise.

1st.—Competition brings into action the most effective system of co-operation among men, and in their final results the two words may be considered synonymous.

2d.—By means of competition the relative share of the product of any given country secured by capital is diminished, while the share of the laborer is increased both absolutely and relatively.

3d.—By means of competition substantial equality in the consumption of the necessaries of life may be attained. As time goes on and abundance increases, the luxuries or comforts of one generation become the necessities and are enjoyed by those which succeed.

4th.—Wages are a consequence or result, and are not a measure of the cost of labor. The better the conditions under which the work is done, the less the cost of a given product measured in terms of labor, and the greater the result or wage measured in terms of money or of what money will buy.

5th.—Civilized man, living under peaceful conditions, increases the means of subsistence by the application of intelligence and skill to all production in a greater ratio than population tends to increase.

6th.—Rent is a tribute rendered by valueless land in proportion to the intelligence, industry, and capital which may be applied to its cultivation, use, or occupancy.

7th.—The burden of general taxation is to be measured by the ratio which the sum of all taxes bears to the net income or savings of the people, rather than by its ratio to the gross product.

8th.—The burthen of a special tax on any given commodity, either foreign or domestic, will be severe or of little moment, according to the subject on which it is imposed. When placed upon an article which enters into the processes of domestic industry, it becomes a great obstruction ; when placed upon an article of voluntary use ready for final consumption, the burden may be small even though the revenue be large, and it is then in exact proportion to the amount of the tax.

9th.—Capital is a force to be applied rather than a substance to be divided. It employs labor and is employed by it—both co-operating of necessity, and not from choice. It follows that the dollars of the fortunes gained in wholesome pursuits are the measure of the services which the owners have rendered to society.

10th.—No acts of legal tender can make two metals circulate permanently at equal values, no matter what adjustments in the weight of coins may be made from time to time. If domestic commerce were not subject to an act of legal tender, it would be conducted on the same basis as that on which foreign commerce is now carried on—namely, by the standard of a given weight of gold.

11th.—The general rate of wages which can be paid in money is made or determined by the sum of money for which the general product can be sold; the less obstruction there is to commerce, either domestic or foreign, the more the general product will bring, the higher the rate of wages will be and the greater the purchasing power of each unit of the wages. Within this limit the rate of wages of each individual is made by himself, and is in the exact ratio of the service which he is capable of rendering to others; it depends upon character, capacity, and industry.

12th.—Insufficient as the product of the United States is compared to what it might be, yet being the result of the cheapest and most effective application of labor and capital yet attained, and being also most free from the burden of destructive taxation, it yields to skill and intelligence the highest rates of wages and the most adequate profits as the necessary result of low cost of production.

Some exceptions have been taken to the propositions submitted in this treatise, while the value of the statistics has been accepted.

It has been said, in one of the most carefully written criticisms, that the so-called law of population propounded by Malthus has been ignored but not disproved. Upon this point no argument will be made; the purpose of the treatise is to present facts and to try to comprehend their meaning. It appears to be a fact, that in this country and in England, during the present century, laborers, as a class, have gained an increasing share of an increasing product; whether such product be considered in ratio to the capital or to the number of laborers engaged upon it. If this be true, then the so-called law of population of Malthus, and the hypothesis that population tends to increase faster than the means of subsistence, are either disproved, or else have been subject to an exception or variation in these two countries, lasting through the whole period since the so-called law was first propounded.

It has also been held that in the distribution of products, whatever the annual product may be, the writer has left no place for rent. It will probably add to the doubt of the capacity of the writer to deal with any thing but statistics, if he expresses a doubt of the existence of rent in the sense in which that word is used by Ricardo and by the later economists of the English school. So far as he has been able to comprehend this theory, it is based mainly upon the varying properties of the soil, subject to modification according to position and to the facilities for marketing its products.

Are there not other modifications or exceptions so numerous as to destroy the apparent rule? Two pieces of land of the same fertility, and, in other respects, each equal to the other, may be so treated that one will yield a large product above the cost of production—that is, will yield rent; while the other will barely yield the cost of production—that is, no rent.

Or the two pieces of land will each yield a large and equal rent; in the one case, being cultivated with the maximum of labor and the minimum of capital; in the other, with the maximum of capital and the minimum of labor.

In neither case do the properties of the soil constitute the measure of the rent.

Is not the rent or income which the soil yields in the long run, over and above the cost of production, chiefly a matter of mental capacity on the part of him who directs its cultivation rather than of its original properties or fertility?

If such be the case, the rents which are attained by virtue of mere possession, and which are claimed by the landlord from the tenant, may not be considered a permanent factor in the distribution of products. The possession of land in England remains the same as it has been, but rents are ceasing to be paid, because the application of science to the mechanism of distribution has almost destroyed the advantage of position of English land. Hence, it may happen, in the course of time, that even English land cannot be made a source of income or rent except to him who applies intelligence and capital directly to its use. In such case, land may be held to be of the same nature as all other instruments of production—that is, as a laboratory which yields product in the exact measure in which capital and labor co-operate in its cultivation. In such case, it may also be asked how rent will differ from any other profit.

Holding this view of the matter, the writer has, therefore, avoided reference to the customary terms used in the distribution of products, and has confined his terms to two shares only,—one share being assigned to the increase of capital, the other to labor.

But even if the share assigned to the increase of capital be divided in the customary way and named in part rent, in part interest, and in part profit, the change in name does not alter the general question.

Is it or is it not true that, as time goes on, the absolute share of the annual product set aside for rent, interest, or profit increases absolutely while it decreases relatively?

Is it or is it not true that the share set aside, assigned to, or earned by labor, according to the common use of that word, is becoming an increasing share of an increasing product; in other words, is the share of the laborer increasing both absolutely and relatively?

So far as the data are to be found, the writer believes in progress *from* poverty, rather than in the assumption of want of equity in the existing system of distribution, which is implied in the phrase "progress *and* poverty."

In this treatise he has made use of the insufficient data within his reach, only with the view of giving a direction to an investigation now admitted to be necessary to the solution of these questions. Witness the increasing importance of, and interest in, the taking of the census, and the establishment of National and State inquiries as to the conditions of labor.

Is there, or is there not, an order in the relations of men to each other, which, when reduced to terms, will constitute the elements of Social Science? If there is, then the historical and statistical method is the only one fit to be adopted, and the *a priori* concepts of many of the accepted economic writers must yield again to the methods of Adam Smith, extended over the wider ground which is now open to him who is capable of occupying it.

To one who has faith, not only in "a power that makes for righteousness," but for human welfare upon this earth as well, the study of these complex problems of modern life may become an absorbing pursuit, no matter how inadequately he may be able to treat them.

In conclusion, the writer may venture to express his gratification with the fact that a second edition of this somewhat disjointed series of economic studies has been called for. He may well be satisfied with the approval indicated by many letters from men of high position, as well as of economists and students of social science; yet the greater satisfaction has consisted in the endorsement of many persons, who, like himself, have been compelled to observe the relations of labor and capital, and to study the forces which make the rate of wages, in the conduct of practical affairs and in the manufacture of goods of many kinds.

There may perhaps be no true or final and satisfactory solution of these complex problems until the members of the unlearned professions of the merchant, the manufacturer, or the underwriter compile the data from their own practical experience for the use of the members of the learned professions who write the books upon social science or teach political economy in the school or in the university. The one may, perhaps, perform the labor, while the other may furnish the mental capital, and from the co-operation of the two the best results may be attained.

Brookline, Feb. 23, 1885.   EDWARD ATKINSON.

SUGGESTIONS TO STATISTICIANS.

In the progress of this work the attention of the writer has been called to the great dearth of what may be called comparative statistics, corresponding to the fifty years' history of cotton factories in all departments : and also to the lack of consecutive statements of the simplest factors in subsistence, like the four years' account of the cost of food for factory operatives. If there are in print statements corresponding to these, the writer would be under a great obligation to any reader who would call his attention to them.

It is by the use of comparative statistics that the relative conditions of working people may be demonstrated, and having ventured to suggest certain methods for rendering our consular reports more complete, a plan is now submitted for comparisons of the condition of laborers at home.

The customary method of treating only the rates of wages and the prices of food, clothing, and rent, is inconclusive, because the proportion of each element in the cost of living varies so much in quantity as well as in value.

May we not, however, establish a *standard* ration,—a *standard* supply of clothing, of fuel, light, and incidentals, and of rent, for certain specified classes of persons whose plane is substantially the same ?

Working from the ration of factory operatives as given in this treatise and from other data, the expenditure might be calculated as follows : of a mechanic in Massachusetts earning $550 to $600 per year, and spending for the necessaries of life $500 per year in supporting a wife and two children, the latter counted as equal to one adult—*i. e.*, a group of three adults corresponding in a measure to the working group of three as shown to exist by the census :

PRO FORMA.	PER YEAR.	PER CENT. OF THE WHOLE COST OF LIVING.
Meat, ¼ lb. fresh to 1 lb. salt, per day, per adult (two children to one adult), 9¼ cents each per day	$100 00	20
Dairy products, ½ pint milk, 1½ oz. to 2 oz. butter, and a scrap of cheese at a fraction under 5 cents per day per adult	50 00	10
Bread, ¼ to ¾ lb. each, at a fraction over 2¼ cents per day per adult	30 00	6
Vegetables, more than one half potatoes, a little over 2 cents each per day	25 00	5
Sugar and syrup, a little less than 2 cents each per day	20 00	4
Tea and coffee, 1 cent each per day	10 00	2
Eggs, ¼ cent each per day	5 00	1
Fruit, green and dry, ¼ cent each per day	5 00	1
Salt, spice, ice, pickles, etc., ¼ cent each per day	5 00	1
Total food	$250 00	50
Fuel and light . . . . . . $28 00 Incidentals, soap, etc. 22 00	50 00	10
Clothing, 35 % cotton ⎫ 45 % woollen ⎬ 20 % sundries ⎭	100 00	20
Rent	100 00	20
Total	$500 00	100 %

This table is only an approximation, and may or may not be a true standard, but it indicates how a very accurate standard can be established. It is given as being suggestive if not conclusive. In some sections the proportions of each element would vary in very considerable measure, and in the same section the proportions may vary in the city and in the country; but would it not be in the power of the Chiefs of the State Bureau of Statistics to establish a fairly accurate *standard*, modelled upon this plan, in respect to three classes of persons in each State:

1. Common laborers, $400 per year, income.
2. Average mechanics, $600 per year, income.
3. Employés of railways or the like whose incomes are about

fifty per cent. higher than those of the average mechanic, or $900 per year.

Each one of these standards being established in each State would serve as a measure for comparing one State with another, and if the average in all States were compiled in one average on each class, a standard would be established for an accurate comparison of the condition of one country as compared to another.

Again : the relative per cent. or proportion of dollars in any given standard, which must be applied to each separate item in the cost of subsistence at the present time, being thus determined, a comparison could be made of the actual condition of laborers in the same State at a much earlier date. For instance, given a family of four persons living upon the total sum of the foregoing table, to wit, $500, it may be assumed that the workman of the same class could spend only two thirds of this sum at some previous date, say in 1840.

Divide the two thirds, or $333.33, in the same proportions that the present expenditure of $500 is divided by. Apply these proportions to the purchase of food, fuel, clothing, and rent at the prices of 1840, and then we have an exact system for the comparison of conditions which do not now exist. We should then be in the possession of the data of wages, prices, and proportionate cost of each of the elements of subsistence.

EXAMPLE.

Suppose wages in 1840 to have been two thirds the present rate. The mechanic now spending $500 per year would then have spent, say $340, in same proportions as he now spends, to wit :

			How much would these sums buy in 1840 ?
Meat	20 %	$68 00	Of beef.
			" mutton.
			" poultry.
			" salt pork.
Dairy	10 %	34 00	" milk.
			" butter.
			" cheese.

				How much would these sums buy in 1840?
Bread	6 %	20 40		Of flour.
Vegetables	5 %	17 00		" potatoes.
Sugar	4 %	13 60		" sugar.
Fuel	6 %	20 40		" coal.
				" wood.
Clothing	20 %	68 00		" printed calico.
				" standard sheeting.
				" 16 oz. cassimere.
				" 4 oz. merino or alpaca.
				" woollen hose.
				" of boots or shoes.
Rent	20 %	68 00		" rooms in a good house.

Having determined quantities in 1840 and compared with the quantities yielded now for the higher wages earned with the same or less labor, we have an absolute comparison of conditions.

The customary comparisons by rates of wages and prices only, fail to meet the case because of the varying proportions expended for meat, bread, sugar, etc., etc. These proportions once established, relative conditions will be easily determined. In preparing this treatise I have been under the necessity of using approximate estimates, because the historical and statistical basis for a true science of wages does not yet exist. The real problem is to determine what the absolute wages in food, fuel, shelter, and clothing now are in this country as compared to others, rather than to determine what the comparative rates of wages in terms of money may be.

Is it true or not that the abundant product of this country yields a larger sum of money to be divided among its workmen than is possible in any other country?

Is it true or not that this sum of money represents a larger supply of the necessaries of life for each dollar expended than in any other country?

If food is cheaper while clothing and shelter are dearer, what are the reasons?

If, with all our advantages of position, of virgin soil, and of freedom from vested wrongs, the laborer cannot earn more and get more for his money in this than in any other land, must we not admit partial failure, and ought we not to proceed at once to correct our methods?

Even since I had prepared these suggestions to statisticians, by the courtesy of Mr. Joseph D. Weeks, of Pittsburg, I have been supplied with the data by which I am enabled to make the following statement regarding the product of a blast furnace, which has been working in the production of pig-iron for the last twenty-five to thirty years.

It is alleged that progress and poverty are correlative terms, and that as the rich grow richer, the poor grow poorer. This is a mere question of fact. Has it been true of iron? There have undoubtedly been very profitable periods during the past thirty years, when the owners of ore beds and coal mines have secured large sums as rent or royalty from those who have worked them. There have also been periods of great profit in the conversion of ores and coal into iron, in which the rich have grown richer. We may not ask, nor expect to be informed, what these profits have been in specific cases; but this we know—that the greater the profit, the more urgent the competition of capital with capital in opening new mines, constructing new furnaces, and producing greater quantities of metal. "Have the poor become poorer?" The main question can be conclusively answered without the disclosure of a single fact of a private nature, and without any inquisition to which any and every capitalist or owner might not cheerfully submit. Witness this statement in regard to iron.

The two periods chosen for comparison are: 1st. 1860 to 1864, inclusive, five years of war, paper-money, inflation, and confusion. 2d. 1875 to 1879, inclusive, the period of slow and steady recovery from a financial debauch, in which the solid and safe specie standard of value was restored.

The furnace which gives the data used in this comparison is one for which all the materials have been purchased at current prices. The data, therefore, give the exact cost of the labor required to convert the coal and ore into iron after they have been delivered.

The furnace is in the eastern part of the country, and is now at a relative disadvantage in procuring material, as compared to some of the establishments in other parts of the country, and its chances of continued success must depend upon the owners overcoming this advantage by skill and intelligence, and by the prompt adoption of easy improvement or labor-saving invention; that is to say, by the sagacious and skilful use of capital.

If we consider the period from 1860 to 1880 historically, it has been one of singular progress in improvements for converting ores into iron, both in the construction of furnaces and in the saving of labor. To whom the benefits of these inventions and improvements have enured, the table shows; but perhaps it may not be amiss to bring the principal changes into more conspicuous contrast, and to compare these changes under the customary classifications:

1st. The margin between the selling price of iron and the cost of materials and labor has decreased $83 \frac{74}{100}$ per cent. The share of the capital has been reduced both absolutely and relatively.

2d. The labor has been rendered less arduous, while the wages of the laborer have been increased $37 \frac{68}{100}$ per cent. The share of the laborer has been increased both absolutely and relatively.

3d. The price of iron to the consumer has been reduced $31 \frac{68}{100}$ per cent.

The measure in money of the gain to laborers is $133 each. For five years' work of seventy-one men, $9,433 per year, $47,215.

The measure in money of the gain to consumers in five years, t $8.87 per ton, is $76,706.

THE LAW OF PROFITS DIMINISHED AND WAGES INCREASED BY COMPETITION, ILLUSTRATED BY THE STATISTICS OF AN IRON FURNACE USED FOR THE CONVERSION OF ORES AND COAL PURCHASED AT MARKET-PRICES, INTO PIG-IRON. PERIODS COMPARED—1860 TO 1864 (FIVE YEARS), 1875 TO 1879 (FIVE YEARS), DESIGNATED RESPECTIVELY I. AND II.

			Increase.	Decrease.
1. Fixed capital	I.	The same in each period.		
	II.			
2. Product of iron, tons	I.	58,959	46 %	
	II.	86,546		
3. Market value per ton	I.	$27 95		31 $\frac{10}{100}$ %
	II.	19 08		
4. Value total product	I.	$1,627,268		
	II.	1,651,298		
5. Cost materials and labor	I.	$1,064,089		
	II.	1,556,889		
6. Per cent. cost materials and labor to value product	I.	65.39	44 $\frac{10}{100}$	
	II.	94.28		
7. Margin for taxes, insurance, cost of selling, incidentals, administration, and profit, if any	I.	34.61		83 $\frac{10}{100}$
	II.	5.72		
8. Sum of wages	I.	$134,214		
	II.	172,491		
9. Hands employed	I.	76		6 $\frac{10}{100}$
	II.	71		
10. Wages per hand per year	I.	$353	37 $\frac{10}{100}$	
	II.	486		
11. Wages per ton	I.	$2 27		14
	II.	1 99		
12. Per cent. of wages to value	I.	8.25	26 $\frac{10}{100}$	
	II.	10.44		
13. Ton product per hand	I.	776	55 $\frac{10}{100}$	
	II.	1,219		

This table might well be named "The indicator of progress *from* poverty of the workman and progress *toward* poverty of the capitalist."

Another graphical method of showing these results is submitted, as follows:

PIG IRON.

Diagram showing the changes which have occurred in a blast furnace used for the conversion of iron ores and coal purchased at market-prices into pig-iron. The conditions of 1860 to 1864 inclusive are taken as a standard, each being called 100, and all represented by the single point at the head of the column on the left; from this point the lines of variation diverge, and the several points in the column on the right show the result of these variations in the averages of product, prices, wages, etc., in 1875 to 1879 inclusive.

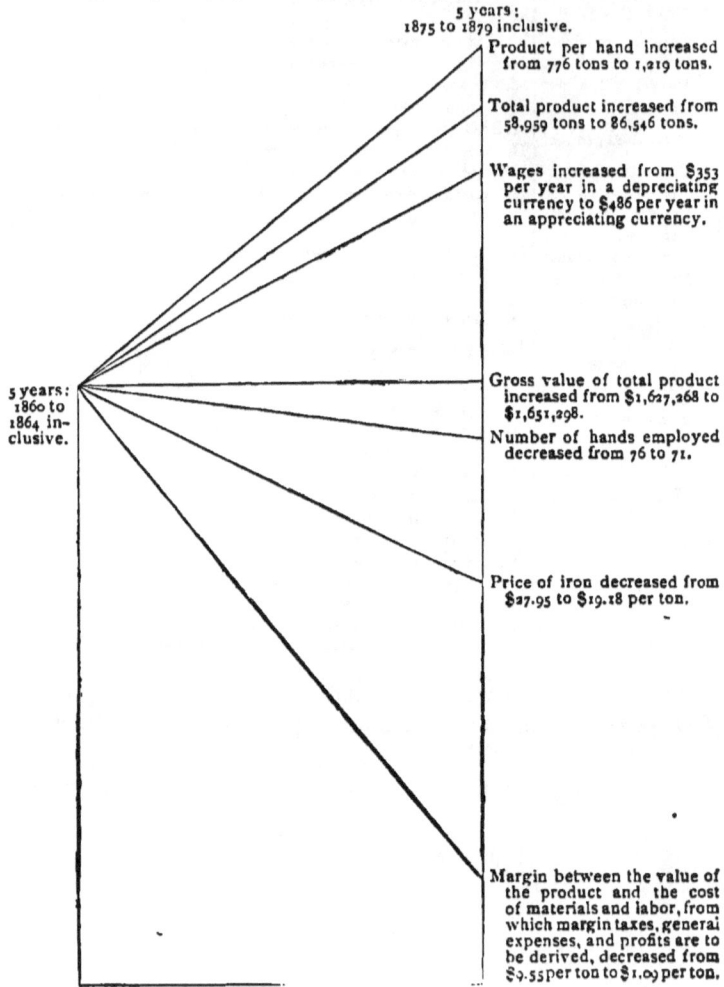

It will be apparent that while the profits of capital may have been much more than ten per cent. in the first period, and must have been much less, if any thing, in the second ; yet such facts can seldom be correctly ascertained, and if given, would not be as useful as to assume a certain uniform rate of profit. It is an absolute rule that if profits rise above a certain rate in any art which is open to free competition, capital will be immediately applied thereto in ample measure so as to bring them down to an average at any given time. If an excess of profit is gained for any considerable period, an excess of capital will be invested, and presently what is commonly called an over-production will occur.

The iron industry has been peculiarly liable to excessive fluctuations, owing to the great fluctuations in the construction of railways, for which so large a part of the product of iron and steel is used.

The attention of statisticians is called to the simplicity of this form. It is merely a digest of the customary annual statements which are made up by all well-conducted corporations or co-partnerships, and any competent accountant could fill up the blanks for any year or series of years. It will be observed that the facts given disclose the progress of the workmen, and the benefit of reduction of price to consumers ; but do *not* disclose the profits of the business in such a way as to be objectionable to owners of works or factories. The diminishing margin between the gross market value of the goods and the combined cost of materials and labor will yet sustain the rule that the profit of manufacturing, of metal work, of transportation, and in fact in all the arts of life, now consists in economy of administration and in saving small fractions in transportation, in the cost of selling, in insurance, taxes, and all the other expenses which of necessity intervene between the primary work of production and the final consumption of all products.

In fact all profit now consists in saving what was once wasted.

It will be apparent to all statisticians that if we can establish the standard ration, the standard supply of clothing, and the standard price of shelter in the way previously suggested, and also secure tables similar to the analyses of cotton fabrics and of pig-iron, the actual progress of working people may be absolutely demonstrated.

It is the purpose of the writer to attempt to procure such data in respect to boots and shoes, hats, paper, cordage, pine lumber, rolled iron, locomotive engines, and many other productions of which the accounts have probably been kept in a uniform way, and for this purpose he will be grateful for any aid which may be rendered.

This is a difficult and uncertain task for an unofficial person to undertake, but even if imperfectly carried out it may yet establish a method which will ultimately lead to exact conclusions.

Finally, in order that all such facts bearing upon the question "What makes the Rate of Wages?" may be brought together for comparison and discussion, the writer invites communications, to be submitted at the meeting of the American Association for the Advancement of Science, in August, in the section devoted to political economy and statistics, of which he has the honor to be chairman. Communications from foreign countries will be gratefully received.

Any persons who are desirous to take part in the collection of such facts, or who will furnish the writer with the requisite data, may address him at No. 31 Milk Street, Boston, Mass., U. S. A.

EDWARD ATKINSON.

BOSTON, *March* 30, 1885.

FORM OF INTERROGATORIES, BY MEANS OF WHICH THE PRICE OF LABOR—I. E., THE RATE OF WAGES, AND THE COST OF LABOR —I. E., THE SUM OF WAGES IN A GIVEN PRODUCT, MAY BE ASCERTAINED.

1. What was a fair valuation of your real estate and machinery at the earliest date from which you can make a consecutive statement of your business ?
2. Beginning at this date, what was the annual product in units, such as pounds of cloth, tons of rails, pairs of boots, etc. ?
3. For each year separately to 1884, inclusive. Or for periods of five years, wide apart ; say, 1856 to 1860, 1866 to 1870, 1879 to 1883—the year 1884 separately ?
4. What was the market-price in each year of some specific unit which has been made of the same kind and same quality, or better, throughout the term ?
5. What was the gross value of the total product each year or each period of five years ?
6. What was the cost of materials and labor combined in each year or period, omitting insurance, taxes, and general expenses, and including as labor—overseers second hands, operators, mechanics, engineers, firemen, and laborers, but *not* including superintendent or clerk ?
7. What was the sum paid for labor as above defined in each year or period ?
8. What was the average number of laborers in each year or each period ?

Consecutive statements from the earliest date to 1884, each year separately, preferred.

Financial years may be given in place of calendar years.

STANDARD COTTON SHEETINGS.

The general tendency of wages toward a maximum and of profits to a minimum is shown by these diverging lines. There have been, of course, great fluctuations but it will be observed that even the reduction in the rate of wages in money, between 1883 and 1885, was accompanied by an increase in the purchasing power of money, so that wages measured in sheetings are now higher than ever before. I believe this is also true if wages are measured in food or woollens. In other words, all who are employed at all can get more for their work than ever before in food, clothing, and shelter.

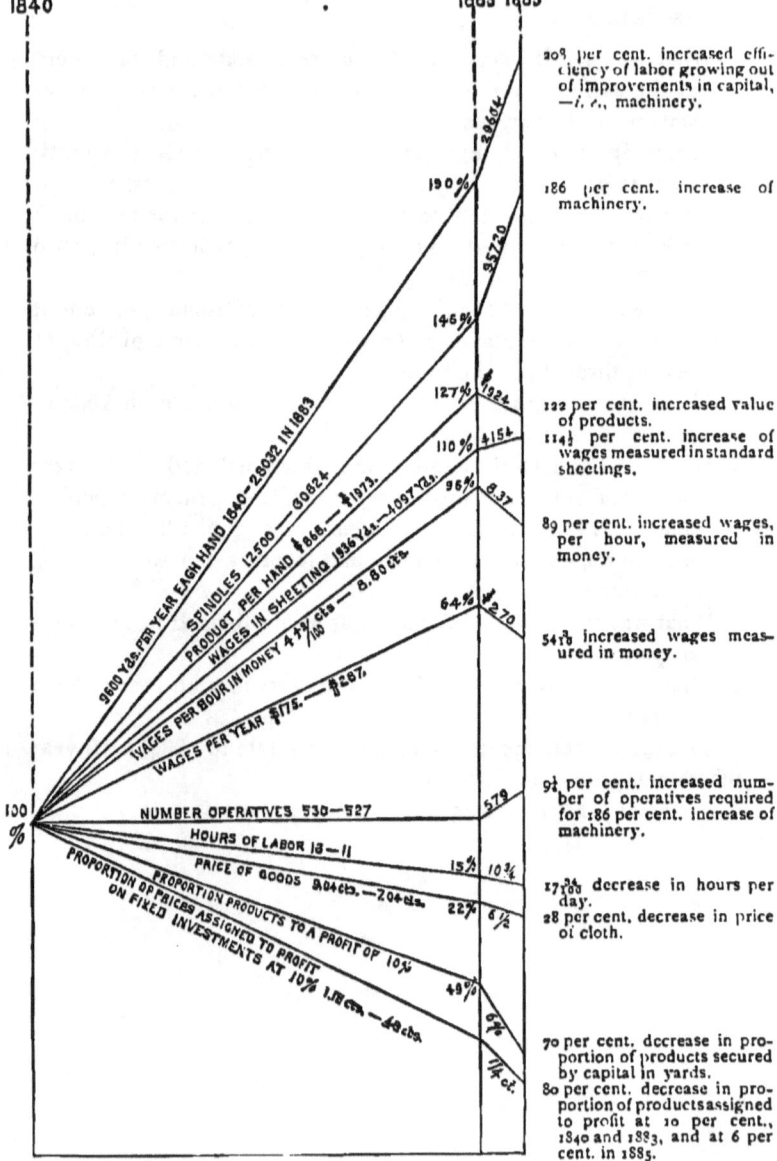

INDEX.

Agriculture in United States, 28, 320, 321
American Association for Advancement of Science, 358
Annual products, 30, 331
Appendices to "Rate of Wages":
I., 91; II., 118; III., 127; IV., 129; V., 139; VI., 158; VII., 171
Appendices to "Railroad, Farmer, and Public": I., 291; II., 298
Arkwright, 79
Armies of Europe, 73
Austria, 16

Bagehot, Walter, quotation from, 185
Balance of trade, 201
Banks and banking, 193
" and manufactures, 217
" State, 222
Banking, elements of, 211
Bastiat, Frederick, proposition from, 23, 89
Beef for prisoners, 163
Beets, 47
Bessemer, Sir Henry, 84
Bi-metallic theory, 200
Bismarck, 13, 16
Blackburn, 343
Blanchard, G. R., 240
Boarding, cost of, 158, 163
Bonanza farms, 76
Brassey, 60
Bread, 75, 167

Bread, analysis of loaf of, 291
Bremen steamer, incident of, 60
Buckle, 67
Bureau of Statistics, 350
Burdens of Europe and America compared, 284

Cairnes' theory of wages, 25
Capital, 25, 188
Carey, Henry C., theory of rent, 340
Cash, 208
Census of United States, 31, 96
" Office Reports, 140
" of Massachusetts, 92
China and India, 69
Cincinnati riots, 270
Cities, growth of, 151
" want in, America and Europe compared, 336
Cities, tendency of poor to collect in, 336
Clothing of various classes, 165
Coal, 78
Coinage of silver dollars, 315
Commercial crises, 314
Competition, 35
" in wages, 12
Conclusions on wage question, 178, 343
Congress, present, 187
Consuls, instructions to, 343
Consumption defined, 199
" in United States, 318

Cotton manufacture, 48, 50, 52, 68
" workers, 42
Corn meal, 161
Cost of living, 338
Crops, amount of land needed for, 335
Cunningham, W., 189

Dairy products, 156
Dakota, 14
Depression, present, 314
Distribution, 9, 346
Dodge, J. R., investigations and diagrams of, 139, 143
Dollars, standard of gold, 319
Duties, 323
Election, result of, 181
Employés in manufactures, 109
" on farms, 111, 320
Employment, lack of, 313
Engel, Dr., investigations of, 134, 166, 169
England, 16, 347
English commerce, 82
" wealth, 136
Exchange, benefit of, 20, 55
" result of, 37

Factories, increase of, 313
" cessation, 313
" operatives, actual consumption of, 318
Factory boarding-house in Massachusetts, 163
Factory boarding-house in Maryland, 158
Fallacies, popular, 26, 58, 62
" counter propositions to, 63
Fare, prisoners', 163
" laborers', 164
Farmers, dependent on foreign market, 305

Fibres, amount transported on railroads, 309
Flour of the West, 75
Food and clothing, 317
Food of workmen and prisoners compared, 169
Form of questions, 359
Formula of production, 48
Frelinghuysen's report, 342
Fuel, amount transported by railway; per year, 309
Fundamental law of labor, 94

Germany, 16, 17, 343
German army, 18
" steamer, incident of, 61
George, Henry, 9, 12, 331
Giffen, Robert, 71, 83
Glasgow, 343
Government, proposed regulation of railroads by, 305
Grain, 22, 232, 235, 309
" crops, table of, 233
Great Britain, land question in, 227
" and her manufactures, 133
Greenback fallacies, 26

"Harmonies of Political Economy," 23
Hen industry, 155
Homespun fabric, 125
Hooper, W. E. & Sons, 158
Howe Bakery, 291
Howe, Samuel, 286

Industry, diversified, 153
Irish Land Acts, 13
Iron, 77, 84, 156, 269, 318, 353, 356
Iron and Steel Association, 240

INDEX.

Kidder, Peabody, & Co., 211

Laborers, 21, 44, 313, 315
Land and labor, need of capital for, 336
Land under national ownership, 332
Law of competition, 118
" exchange, 197
Legal-tender Acts, 197, 305
" United States notes, 221
Louisiana purchase, 203

Machinery and agriculture, 99
" effect of, in manufacture, 35.
Malthus, 15, 22, 339, 345
Mauger & Avery, 240
Mansfield, Judge, 4
Manufactures, mechanics, and mining, 306
Memorial Hall, system of, 164
Metals, amount transported on railways, 309
Metaphysics of exchange, 189
Mexican dollars, 202
Middle States, manufacture and commerce of, 306
Mid-Lothian, 343
Miners, 329
Minnesota, 15
Money, 26
" definition of, 28, 194, 220
" false and true, 207, 211, 221, 223
Money, fiat, 4, 135, 194
" paper, 227
Montana, improvement of, 22

National Bank, 205
" " work of, 216
" Legislature, 287

National revenue, collection of, 305, 320
Necessities of life, exchange of, 312
Newcastle, 343
New England, manufactures of, 306
Nimmo, Joseph, Jr., report of, 139, 341
North Carolina and New England, 68
Nutritious food, 164

Occupations, 104, 106, 305, 320
" summary of, 149, 310
Ohio, 147, 325
Oldham, 343
Oregon, 152
Over-production, 55, 182

Parsons, Judge, 4
People of United States, expenditures, 337
Pepper, 202
Persons engaged in production and distribution, 277
Persons affected by foreign markets, 322
Phillips, Wendell, 12
" Pillar " dollars, 202
Political economy, French system of, 23
" Poor's Railway Manual," 240, 341
Population of globe, 22
" United States, 26, 28, 315
President of United States, result of choice, 181
Product, annual, in United States, 26, 95, 353
Product, division of, 70
Production defined, 9, 10
Products, tables of, 156, 317, 327
Professionals, 305

Profits of capital, 357
"Progress and Poverty," 9
Progress in working iron, 354
" of United States, 83
Protection of domestic industry, 320

Railroads, 27, 268
" adjustment of value of stock, 263
Railroads, capital, relation to farms and factories, 256
Railroads, change wrought by, 231
" charges, 292
" " effect on cost of meat, 295
Railroads, construction of, 43, 185, 315.
Railroads, diagrams, 274
" Farmer, and Public, 101, 175, 231
Railroads, freight charges, 235, 252
" local traffic, 252, 306
" New York Central, 75
" mileage, 240, 252, 257, 316
" government regulations, 294
Railroads in Ohio, 148, 261
Railroads unnecessary, 260
" watered stock of, 239
Railway Manual, Poor's, 237
" lands, large sale of, 314
" officials and laborers, 107, 313
" panics, 185
" " and commercial paralysis, 254
Railway rates higher with small traffic, 309
Railway service, extension of, 99, 305
" system, 99
Raw material, 323
Reconstruction, 182

Reign of terror, 16
Religious dogma, 19
Rent, formulas of, 340
Reports of Railway Commissioners, 341
Resources of United States, 74
Resumption Act, 226, 233
Ricardian theory of rent, 340
Rogers, J. E. Thorold, 189
Russia and nihilism, 16

Sabine, H., 240
Science, advancement, 21
Ship-building on the Delaware, 330
Silver Act, 154, 305
" compared with other products, 318
Silver dollars, uncertain standard of value, 316
Slater, Samuel, 80
Slavery, abolition of, 99
Smith, Adam, 348
Southern States, agriculture of, 306
Speculation, 225
Spinning-jenny, 84
Standing army, 281
Statisticians, suggestions to, 349
State banks, 222
Suez Canal, 19
Sugar, 161
Summary of wealth in United States, 117
Surplus revenue in United States, 57
Tables :
Average work and wages, 118-121, 127
Averages of wages, 129
Consumption of food, 159, 160, 162, 163, 165, 171, 172, 175, 350, 351
Grain crops in United States, 233, 296

INDEX.

Law of profits, 355
Occupations, 149-153, 305, 310 *et seq.*, 325, 326, 328
Only approximately accurate, 313
Products, 157
Products, value of, compared to manufactures, 140
Railway charges, 293
" construction, 274
" employés, 277
Railways, farms, and manufactures compared, 256
Railroad mileage, 234, 241, 242, 244-252
Railway traffic, 148
Railways, tons moved by, 262
Relative burdens of Europeans and Americans, 284
Relative taxation, 154

Tariff, 82, 325
Taxes, 29, 33, 99, 180, 324
" difference between, on raw material and finished products, 323, 324
Temple, Sir Richard, 74
Theories of wages: Thornton, Cairnes, Walker, 24
Timber, amount transported by rail, 309
Trade and transportation, 306
Traffic, relation of volume to rate of charge, 307

Unemployed, work for the, 279

United States, land in, 332
United States, area of, 334
" resources of, 74, 334
" Supreme Court of, 4

Value of manufactured goods, 165
Vanderbilt, Cornelius, 38
Vegetable food, 161
Vienna under martial law, 16

Wages, rate of, 9, 341
" subdivision of, 39
" question, 62
" word defined, 70
" high rate of, in United States, 74
Wages, statistics, 130
" theory of, 24
Walker, E. H., 240, 341
" Francis A., 24
War, dangers of modern, 19
" *versus* work, 72
Wealth of United States, 31, 96
" secret of, 136
Weeks, Joseph D., 353
Western States, grain in, 306
Wheat, 14, 296
Williams, consul at Rouen, report of, 343
Wool, 156, 266, 326
Wright, Carroll D., 129, 166, 169, 341

Yorkshire, 343

www.ingramcontent.com/pod-product-compliance
Lightning Source LLC
Chambersburg PA
CBHW032042220426
43664CB00008B/828